BIRD BRAIN:
AN EXPLORATION OF AVIAN INTELLIGENCE

BIRD BRAIN:
AN EXPLORATION OF AVIAN INTELLIGENCE

DR NATHAN EMERY

WITH A FOREWORD BY FRANS DE WAAL

PRINCETON UNIVERSITY PRESS

PRINCETON AND OXFORD

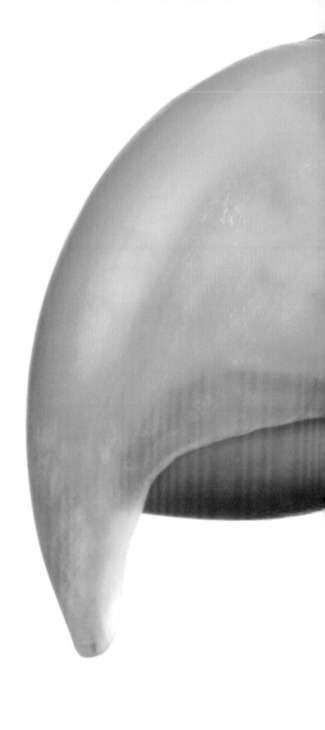

This edition published in the United States of America
and Canada in 2016 by Princeton University Press
41 William Street
Princeton, New Jersey 08540
press.princeton.edu

British Library Cataloguing-in-Publication Data
A catalogue record for this book is available from the
British Library

Library of Congress Control Number: 2016932443

ISBN: 978-0-691-16517-2

This book was conceived, designed, and produced by
Ivy Press
Publisher Susan Kelly
Creative Director Michael Whitehead
Editorial Director Tom Kitch
Art Director Wayne Blades
Project Editor Jamie Pumfrey
Commissioning Editor Jacqui Sayers
Designer Simon Goggin
Illustration Concepts Nathan Emery
Illustrators John Woodcock, Jenny Proudfoot & Kate Osborne
Picture Researcher Alison Stevens

Printed in China

10 9 8 7 6 5 4 3 2 1

Front cover image Getty Images/Mint Images—Art Wolfe
Back cover image Shutterstock/Dejan Stanic Micko

CONTENTS

FROM BIRDBRAIN TO FEATHERED APE

Although "bird brain" is used as a term for stupidity, we now know that some birds have brains that process information in similar ways to the great apes. We investigate how our understanding of the avian brain and intelligence has evolved and what this might tell us about the "clever club" of animals.

WHERE DID I HIDE THAT WORM?

Birds are masters of navigation. They are the furthest animal travellers, found in every habitat on Earth, and some have a memory that would outperform the greatest human mnemonic, remembering the location of tens of thousands of different food caches. We investigate the sophisticated spatial memory and migratory sense of birds.

GETTING THE MESSAGE ACROSS

Communication is at the heart of all bird behaviour; from the colourful displays of attraction and courtship, to singing to defend a territory and calling to warn others about a predator. We look at how these signals exploit the complex visual and auditory senses of birds and present parallels with human language and non-verbal communication.

FOREWORD

Despite a long tradition of research on bird navigation, imprinting in ducklings, and song learning, scientists have always carefully avoided the term "cognition" when it came to birds. This was partly since any contemplation of what might be going on in the heads of animals was taboo, but also due to the special anatomy of the avian brain. Their central nervous system was thought to lack anything like a prefrontal cortex. Having feathers, it was concluded, simply couldn't go together with advanced learning, let alone thinking. That the prototypical laboratory bird—the pigeon—has a tiny brain didn't help much either. Birds were lumped in with fish and insects as organisms driven by instinct.

That we now think differently about bird intelligence is in no small measure thanks to pioneering research that started as a trickle in the 1990s, which by now has grown into a steady stream. Complex cognitive concepts, such as planning for the future or theory-of-mind, were translated into carefully controlled tests. The results have been eye-opening, and hard to disavow by skeptics due to the rigor of the experiments. This work has revolutionized our appreciation of what birds are capable of. Nathan Emery has been at the forefront of this research, while emphasizing convergent evolution to explain the similarities with primates and other large-brained mammals. Species don't need to be related to accomplish similar cognitive feats. We used to think in terms of a linear ladder of intelligence with humans on top, but nowadays we realize it is more like a bush with lots of different branches, in which each species evolves the mental powers it needs to survive. As a result, some bird species may be mentally closer to the primates than anyone imagined.

One of the first examples was Alex, the African grey capable of verbally labelling objects. The parrot would face a tray full of different items, feeling every single one of them with his beak and tongue. After this exploration, he would be asked what the blue object is made of. By correctly answering "wool," he'd combine his knowledge of color and material with his memory of what this particular item felt like. Even though the claim never was that Alex possessed language, he answered questions that no one thought could be answered without language.

Equally remarkable was Betty the crow, who manufactured a tool out of straight metal wire by bending it into a hook. This way, she could fish a little bucket with food out of a tube, something impossible with an unbent wire. As a New Caledonian crow, Betty was a manufacturer of tools, like her species in the wild. There were also the Western scrub jays, which seemed to grasp what others knew. Caching food, such as mealworms, is a natural behavior in these birds, but the tactic is vulnerable to pilfering. This may explain why, if they know that another jay has been watching them, jays will quickly re-cache their worms as soon as they are alone. It is as if they realize that the other knows too much.

I myself played hide-and-seek games with my tame jackdaws, which I kept as a student until they flew away. These particular corvids were also the favorites of early ethologists, who described their behavior in delightful detail, yet rarely mentioned their intelligence. The latter topic was (and is) surrounded by heated debates dominated by the proposal that everything boils down to associative learning. For the longest time, this was the fallback position. Only when an avalanche of experiments demonstrated the flaws of explanations based on reward and punishment, did cognitive claims gain the upper hand. Animals, including birds, sometimes solve problems that they have never tackled before, demonstrating an immediate "insightful" grasp of the contingencies.

As a result, scientists are not reluctant anymore to propose thought processes in animals. Birds show evidence for precise memory of past events, perspective-taking, prospective planning, versatile tool use, reconciliation, and empathy. While each study in and of itself might not have changed our perception, the accumulation of knowledge, so nicely captured

here in a single book with elegant drawings and photographs, strongly favors a significant upgrading of what the avian mind is capable of.

One indication comes from the bird brain itself. There are about 10,000 bird species, which show enormous variation in brain size. Given how "expensive" brain tissue is (it needs about twenty times more energy per unit than muscle tissue), there must be excellent evolutionary reasons why certain families have invested in large brains. Evolution generally doesn't produce surplus capacity. Since corvids and parrots are endowed with brains roughly the size of those of primates, after correction for body size, we should not be surprised that they possess mental powers similar to those of monkeys, perhaps even apes. Furthermore, we now know that the structure of avian brains was misjudged in the past. The forebrain derives from the pallium, which also produced the mammalian neocortex. This makes the bird brain structurally far more similar to that of mammals than previously thought.

The result that we here hold in our hands is a fascinating, visually attractive, smoothly written summary by one of the pioneers of the up and coming field of avian cognition. You'll find no better guide to the latest knowledge and the current debates. That birds are only distantly related to the primates, including us, is surely not something we should hold against them. As Emery explains, many of the cognitive capacities that we observe in birds make perfect sense given their natural history and the challenges encountered in nature. Having feathers, as it turns out, can very well go together with a sophisticated mind that tackles problems with great flexibility and resourcefulness.

Frans de Waal
C. H. Candler Professor of Psychology and Director of the Living Links Center, Emory University, Atlanta, USA

Why birds?

Birds have likely fascinated us since we became a species. We desire their ability to fly and latterly have been able to use our intelligence to match and even surpass the life of birds among the clouds. However, we have never envied their intelligence!

Birds' cognitive abilities have a terrible reputation; we even use "birdbrain" as a term for stupidity. But are we being fair when we characterize them as dumb? They are certainly very adaptable animals, found across the world in extremes of temperature, climatic variability, and habitat, including the coldest parts of the Antarctic and the hottest deserts. They are also a diverse taxon with more than 10,000 known species. Some are very successful with millions of individuals, colonizing large tracts of the planet and often sharing our own habitats.

What of their intelligence? Can we say that the lowliest chicken is clever? Well, it depends on what we mean by intelligence, a subject we'll address in the next section. If we apply a common use of the term, we automatically differentiate birds into either dullards or geniuses based on their role in our cultural history. But is this fair?

Although pigeons do not, at first, appear to be avian Einsteins, some of their abilities are keenly attuned to the challenges they face, especially the business of finding food. A species that specializes in finding and eating grain, pigeons have to locate very small food items hidden within a substrate that looks very similar to the food. Watch closely the next time you observe pigeons eating in the park. Foraging requires the pigeon not only to have an image of the target food but to be able to discriminate this image from other slightly dissimilar images, such as the texture of the floor. Pigeons have shown that they can tell apart all sorts of visual stimuli, an ability likely to be related to their food-finding skills. But does this make them clever? The ability to learn quickly, and especially to be flexible in learning, is linked to intelligence. However, despite pigeons' considerable discrimination skills, learning something quickly—in, say, one or two trials—is not one of them. Pigeons tend to require hundreds of trials to learn something new, so pigeons are not perhaps the best place to look for avian intelligence. Not that a lot of work has been done in this field as, surprisingly, out of 10,000 species of birds very few have been tested for their intelligence. However, two main focus groups are the corvids (the crow family) and the parrots.

Corvids have long been considered clever birds. They are the stars of many myths and legends from the raven creator myths of Native Americans to the Viking god Odin's two ravens, Hugin and Munin (Thought and Memory), sitting one on each shoulder bringing him news each day from across the world. Legend has it that if the resident ravens ever leave the Tower of London, Britain will fall (though recent research has revealed that ravens were probably introduced no earlier than the Victorian era). There is scarcely a Hollywood horror, fantasy, or suspense movie that does not incorporate crows as symbols of death, disease, witchcraft, or woe, with Hitchcock's *The Birds* being the classic example.

On the other hand, similar myths do not exist for parrots, whose reputation has arisen from their capacity for imitating human speech. Parrots were originally collected by the European aristocracy for their beautiful plumage, but they rapidly gained prominence once it was discovered that they could be trained to talk.

Birds in general are interesting from a scientific standpoint, as their brains that have taken a different evolutionary path to that of mammals while at the same time, in many cases, arriving at what seem to be similar solutions to the same problems.

Above Crows and ravens are frequently associated with death, disease, and horror in literature and movies. The classic example is Hitchcock's movie *The Birds*, where hordes of demonic birds terrorize a quiet Northern Californian seaside community.

"The poor development in birds of any brain structure corresponding to the cerebral cortex of mammals led to the assumption among neurologists not only that birds are primarily creatures of instinct, but also that they are very little endowed with the ability to learn. There is no doubt that this preconceived notion, based on a misconceived view of brain mechanisms, hindered the development of experimental studies of bird learning."

William Thorpe (1963)

Learning & Instinct in Animals (2nd edition)

Above The Norse god, Odin, had two pet ravens, Hugin (Thought) and Munin (Memory), which he sent out every morning to bring back news from across the world.

WHAT IS INTELLIGENCE?

What do we mean when we say that an animal is intelligent? Scientists mean something specific by intelligence, especially in creatures without language: the ability to flexibly solve novel problems using cognition rather than mere learning and instinct.

Intelligence in action is the application of cognition outside of the context in which it evolved. An animal may have evolved a specific skill that enables it to deal with a particular ecological problem, such as predicting the behavior of group members or distinguishing large from small quantities, but it cannot use these same skills to address different problems for which the skills did not evolve. However, the flexibility to be able to transfer those skills is probably what distinguishes intelligent from cognitive species.

Cognition refers to the processing, storage, and retention of information across different contexts. In the wild, birds use cognition to process information, enabling them to survive but not necessarily to solve problems. A pigeon that distinguishes foods from non-foods does not need to stretch its mental muscles as much as a crow that creates and modifies a tool to reach a grub hidden inside a tree trunk, fashioning the tool to the correct length in order to reach the treat.

Both are challenges related to procuring food, but one requires a wider range of skills than the other.

One important consideration is that intelligence is not a mechanism. A specific behavior can be perceived as intelligent based on its outcome—such as the solving of a problem—but that does not mean that this solution is achieved using similar processes to those used by a human. The animal may employ sophisticated cognitive processes—perhaps using imagination (thinking about objects, events, and actions not currently available to perception), or forward planning (prospection), or requiring an understanding of how events (actions) are related to their consequences (causal reasoning)—and these cognitive acts may be variously deployed in different contexts. But they may also be the result of trial-and-error learning (learning the best course of action after repeated experiences of the same event) or simpler cognitive processes for which that particular species has evolved a solution. The specific mechanisms underlying animal behavior are frequently the object of controversy and debate, especially in creatures more distantly related to us. This book attempts to present different perspectives on what may underlie seemingly intelligent bird behavior: from instinct, learning, and cognition to imagination, forethought, and insight.

Left Merlina is one of the ravens at the Tower of London. She has formed a strong bond with Chris Skaife, the Raven master, but also likes carrying around sticks and even plays dead to the delight of the crowds who come to see her antics.

Right This murmuration of starlings seems to be behaving entirely reflexively. A predator appears and the individuals on the edge of the flock try to escape, with the birds in the center following them blindly. In this case, it is probably the better strategy for survival to switch from thinking to gut instinct.

The evolution of avian intelligence

Not all birds were created equal. The term "birdbrain" remains apt for many species. Consider the case of the dodo, the archetypal dim-witted bird.

The species lived in splendid isolation on the island of Mauritius in the Indian Ocean until contact with European sailors in the seventeenth century led to its extinction in just a few decades. Although the relatives of dodos (pigeons and doves) are not thought of as the smartest of birds, can we put the dodo's demise down to its own stupidity? Certainly, having no natural predators and not having had much contact with humans before the seventeenth century, they had little or no reason to fear us. If dodos had had the capacity for rapid learning, perhaps they might have adapted quickly and learned to escape their human hunters, but they were up against the most efficient and effective killer the

Evolutionary Tree

This evolutionary tree visualizes the evolution of mammals, reptiles, and birds from a common ancestor—a stem amniote. Despite their reputation for stupidity, birds are not an ancient group. Indeed, compared to mammals they are modern, being the most recently evolved. Birds are so closely related to dinosaurs that they are classified as avian dinosaurs.

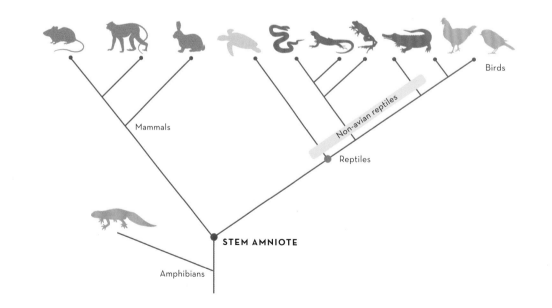

Mammals

Non-avian reptiles

Reptiles

Birds

STEM AMNIOTE

Amphibians

planet has ever seen. Given the dodo's clumsy body design—large and flightless—and that it had nowhere to run, it's clear that dodos were in the wrong place at the wrong time, though being stupid didn't help!

More than 50 percent of birds are members of the songbird family or passerines. In fact, most of the birds we encounter every day in our gardens and parks are passerines, including sparrows, thrushes, finches, titmice, robins, blackbirds, and crows. Although not all members of this family are melodious singers, as anyone who has experienced the loud cawing of a crow will testify, all learn vocalizations specific to their species and, indeed, have evolved a special brain circuit to do so. This ability, rare in the animal kingdom, shares properties with human language which will be examined in Chapter 3.

Although birds have been studied with respect to the structure and function of their brains, their learning, and cognition for over a century, very little is known about the cognitive abilities of more than a tiny proportion of species. Most species are not kept in laboratories and thus are unavailable for experimental study, so our best ideas about their intelligence are only guesses based on their relative

Above left and above Despite there being almost 10,000 species of birds, only a few have yet to be studied for their cognitive abilities. Some, based on their lifestyles and relative brain size, such as this woodpecker (left), hornbill, and falcon (right), are likely to also demonstrate smart behavior in intelligence tests.

brain size (in comparison to their body size; see Chapter 1), their diet, social system, habitat, and life history (how long the species lives and how long the young take to develop to independence). These clues help build a picture of what these species may need their brains for—finding food, relating to others, building a home—but without being able to run experiments the picture can only be a sketch. Nonetheless, this technique is still useful for making predictions as to how intelligence may have evolved, specifically in those species we would expect to be the intellectual heavyweights. Three groups of birds— woodpeckers, hornbills, and falcons—possess some or all of the traits displayed by species known to be smart (The Clever Club; Chapter 1) but have yet to be tested. All three groups are outside the passerines but are closely related, so any cognitive skills they may have are likely to have evolved independently (that is, not from a common ancestor).

1 FROM BIRDBRAIN
TO FEATHERED APE

Evolving view of the avian brain

The erroneous idea of the birdbrain—that birds are not endowed with intelligence—can be traced to the latter part of the nineteenth century and the work of comparative anatomist Ludwig Edinger.

EDINGER'S ERROR

In an encyclopedia of animal brains, Edinger claimed that the avian brain was composed largely of the striatum, the part of the brain responsible for instinctual or species-typical behavior, with little or no areas responsible for thinking, such as the cortex. Edinger's logic was that if the avian brain had evolved from the striatum, then birds should not be capable of thought. Edinger's idea held sway well into the twentieth century despite studies on bird intelligence suggesting that their behavior was anything but driven by instinct alone.

BIRD LEARNING COMES OF AGE

Studies of complex learning throughout the 1950s and 1960s, including song, imprinting, and imitation, were to change opinions about the mental capabilities of birds. At this time, the idea that animals could initiate their own actions, solve problems, and think was completely contrary to the long-held assumption that they were automatons and only responded to changes in their environment based on positive (rewarding) or negative (punishing) consequences. In the 1970s, a new wave of research on birds focused on adaptive specializations or cognitive abilities that evolved to solve specific environmental or social challenges. For example, many birds hide food for future consumption (caching). To locate hidden caches, especially when spread over a wide area a long time in the past, requires a sophisticated spatial memory. Birds that are most dependent on caching have a better spatial memory than those that don't cache or who are less reliant on cached food. This cognitive ability appears to have evolved with an increase in the size of the specific brain region associated with remembering location, the hippocampus (see Chapter 2).

Left A Florida scrub-jay uses its memory to search for a food cache it hid some time earlier. It needs to remember where it hid the food, but also how long ago to make sure the food is still edible and take into account who or what might have been watching when they hid it.

A NEW THINKING

The 1990s saw a flurry of interesting studies on avian behaviors thought to be uniquely human or only seen in great apes. Gavin Hunt found that New Caledonian crows made two different types of tools—Pandanus leaf and hook stick—that were used for different tasks. Irene Pepperberg revealed previously unheard-of linguistic abilities in a language-trained African grey parrot called Alex. Nicky Clayton and Tony Dickinson developed a method based on caching to discover that Western scrub jays thought about specific past events, so-called episodic-like memory.

In parallel to the exciting findings in avian cognition were findings from avian neuroscience. Bird brains were found to do things not seen in mammalian brains that could explain how birds could achieve identifiable cognitive feats with brains much smaller than mammals. Bird brains could support multitasking, with one hemisphere controlling one behavior (such as looking out for predators) while the other hemisphere controlled a different behavior simultaneously (such as looking for food). Adult brains could produce new neurons (neurogenesis)—either seasonally, as in the case of the hippocampus or song control system, or when needed, such as remembering caching events.

Edinger's earlier ideas on the avian brain were questioned by studies on neuroanatomy, neurochemistry, evolution, and development with the result that in 2004 a complete change was made to the naming of the parts of the avian brain, reflecting a new understanding of how it had evolved. No longer was the avian forebrain seen as consisting of the striatum; rather the forebrain evolved from a pallium shared with ancestral reptilian and mammalian cousins. These new findings placed the new study of avian cognition on a strong foundation—so much so that more recent findings suggest the term "birdbrain" should now be used as a compliment not an insult!

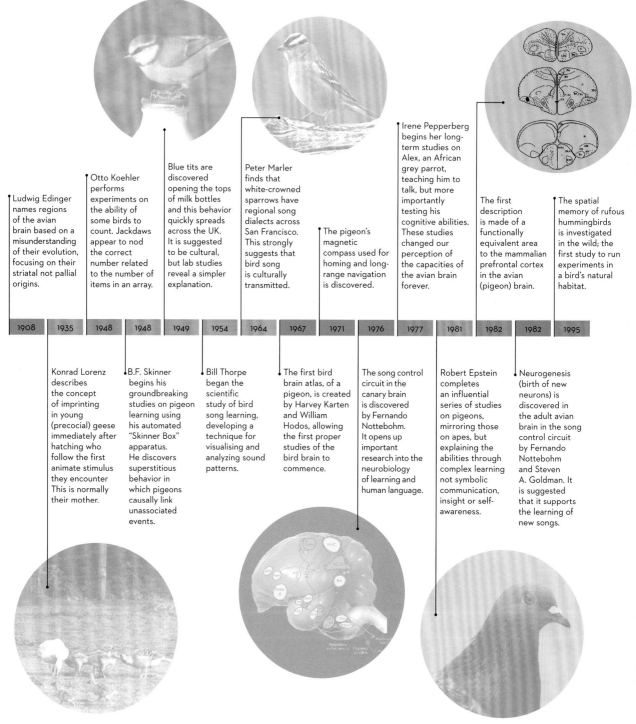

KEY FINDINGS IN THE HISTORY OF AVIAN BRAIN, LEARNING, AND COGNITION

Date	key event in studying avian brains
Date	key event in studying avian learning
Date	key event in studying avian cognition

Ludwig Edinger names regions of the avian brain based on a misunderstanding of their evolution, focusing on their striatal not pallial origins.

Otto Koehler performs experiments on the ability of some birds to count. Jackdaws appear to nod the correct number related to the number of items in an array.

Blue tits are discovered opening the tops of milk bottles and this behavior quickly spreads across the UK. It is suggested to be cultural, but lab studies reveal a simpler explanation.

Peter Marler finds that white-crowned sparrows have regional song dialects across San Francisco. This strongly suggests that bird song is culturally transmitted.

The **pigeon's magnetic compass** used for homing and long-range navigation is discovered.

Irene Pepperberg begins her long-term studies on Alex, an African grey parrot, teaching him to talk, but more importantly testing his cognitive abilities. These studies changed our perception of the capacities of the avian brain forever.

The first description is made of a functionally equivalent area to the mammalian prefrontal cortex in the avian (pigeon) brain.

The spatial memory of rufous hummingbirds is investigated in the wild; the first study to run experiments in a bird's natural habitat.

| 1908 | 1935 | 1948 | 1948 | 1949 | 1954 | 1964 | 1967 | 1971 | 1976 | 1977 | 1981 | 1982 | 1982 | 1995 |

Konrad Lorenz describes the concept of imprinting in young (precocial) geese immediately after hatching who follow the first animate stimulus they encounter This is normally their mother.

B.F. Skinner begins his groundbreaking studies on pigeon learning using his automated "Skinner Box" apparatus. He discovers superstitious behavior in which pigeons causally link unassociated events.

Bill Thorpe began the scientific study of bird song learning, developing a technique for visualising and analyzing sound patterns.

The first bird brain atlas, of a pigeon, is created by Harvey Karten and William Hodos, allowing the first proper studies of the bird brain to commence.

The song control circuit in the canary brain is discovered by Fernando Nottebohm. It opens up important research into the neurobiology of learning and human language.

Robert Epstein completes an influential series of studies on pigeons, mirroring those on apes, but explaining the abilities through complex learning not symbolic communication, insight or self-awareness.

Neurogenesis (birth of new neurons) is discovered in the adult avian brain in the song control circuit by Fernando Nottebohm and Steven A. Goldman. It is suggested that it supports the learning of new songs.

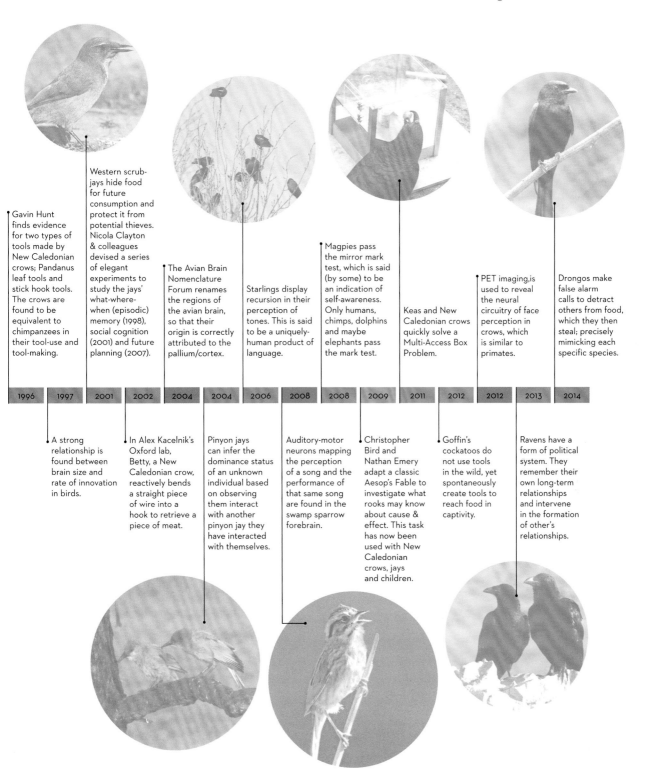

Gavin Hunt finds evidence for two types of tools made by New Caledonian crows; Pandanus leaf tools and stick hook tools. The crows are found to be equivalent to chimpanzees in their tool-use and tool-making.

Western scrub-jays hide food for future consumption and protect it from potential thieves. Nicola Clayton & colleagues devised a series of elegant experiments to study the jays' what-where-when (episodic) memory (1998), social cognition (2001) and future planning (2007).

The Avian Brain Nomenclature Forum renames the regions of the avian brain, so that their origin is correctly attributed to the pallium/cortex.

Starlings display recursion in their perception of tones. This is said to be a uniquely-human product of language.

Magpies pass the mirror mark test, which is said (by some) to be an indication of self-awareness. Only humans, chimps, dolphins and maybe elephants pass the mark test.

Keas and New Caledonian crows quickly solve a Multi-Access Box Problem.

PET imaging,is used to reveal the neural circuitry of face perception in crows, which is similar to primates.

Drongos make false alarm calls to detract others from food, which they then steal; precisely mimicking each specific species.

| 1996 | 1997 | 2001 | 2002 | 2004 | 2004 | 2006 | 2008 | 2008 | 2009 | 2011 | 2012 | 2012 | 2013 | 2014 |

A strong relationship is found between brain size and rate of innovation in birds.

In Alex Kacelnik's Oxford lab, Betty, a New Caledonian crow, reactively bends a straight piece of wire into a hook to retrieve a piece of meat.

Pinyon jays can infer the dominance status of an unknown individual based on observing them interact with another pinyon jay they have interacted with themselves.

Auditory-motor neurons mapping the perception of a song and the performance of that same song are found in the swamp sparrow forebrain.

Christopher Bird and Nathan Emery adapt a classic Aesop's Fable to investigate what rooks may know about cause & effect. This task has now been used with New Caledonian crows, jays and children.

Goffin's cockatoos do not use tools in the wild, yet spontaneously create tools to reach food in captivity.

Ravens have a form of political system. They remember their own long-term relationships and intervene in the formation of other's relationships.

What do bird brains do?

Brains are not just for thinking. Most animals probably do not have the capacity for thought, and so their brains are used for more mundane things like keeping the body ticking over, moving it around, and reacting to stimuli in their environment.

WHAT BRAINS DO

Brains are a means of interacting with the world, receiving information from different senses—sight, hearing, smell, touch, and taste—which is then interpreted by being compared with stored representations of information, either held in memory, such as when recognizing a familiar face, or learned through repeated experiences, such as knowing how to ride a bicycle. Brains make decisions about how to react to information, ignoring most of what is received, only attending to that which is (biologically) relevant. Decisions are made in the context in which information is currently experienced, as well as based on previous experiences, on current motivational or emotional states, and on social context. Once a decision has been made, an action plan is initiated leading to a specific behavior.

WHAT BIRD BRAINS DO

For birds, this decision may be how to react to a vocalization. If this is an alarm call (presented by a reliable, nearby source), then the appropriate response is to move quickly in the opposite direction to the call, namely away from a predator. If a potential mate

Left Birds such as the parakeet fly at high speeds in a complex three-dimensional world and as a result the avian brain must react much quicker to avoid colliding with the canopied forest.

produces the call, inviting sexual behavior, then an appropriate response would be to move closer to the source of the call. In either case, the brain interprets the content of the information (good or bad) and directs the body to act appropriately (approach or retreat).

THE 100MPH BRAIN

Birds often make decisions at much quicker speeds than mammals because of the environments they occupy and the vagaries of their lives, often living at high speed in flight. By contrast, a typical mammal, such as a rat, scurries around its environment but is dependent on smell not sight. Rats don't travel quickly and don't need to make rapid decisions. Monkeys make faster decisions as they travel rapidly through a more crowded environment, often swinging through the trees or being chased by a predator. Primates rely predominantly on sight and sound, both rapid routes for communicating information. Yet, primates do not have to process information as rapidly as a typical flying bird living in a complex three-dimensional world, flooded with color, and sources of danger and information. Birds seem to make these decisions with ease, yet the questions remain: how do they do it and what parts of their brains are used to process this information? Is there something about how the avian brain is wired up that helps birds process information more rapidly than other creatures?

Comparing Brain Sections

The right hemisphere of an owl (left), rat (center), and monkey (right) highlights two important regions shared by birds and mammals, the hippocampus (gray) and cortex (red).

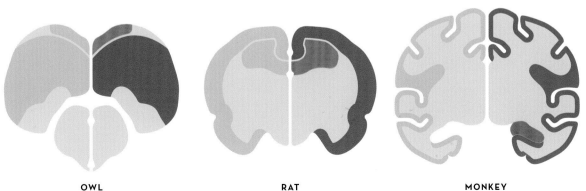

OWL   RAT   MONKEY

The avian brain

Despite more than one hundred years of study, we know very little about the structure and function of the avian brain. This is largely because birds were thought unimportant and uninteresting, especially when studying the neural basis of cognition.

THE AVIAN BRAIN IN A NUTSHELL

We now know that the avian pallium has more in common with the mammalian neocortex than previously assumed, because they both evolved from the pallium of a stem amniote ancestor more than 300 million years ago. What we know about the avian brain is restricted to a few species: the pigeon, the domestic chick, and a few songbirds, such as the zebra finch. None of these species figure in the list of the world's smartest birds, so our current knowledge is far from complete, because there are approximately 10,000 species of birds, all with different brain architectures.

All vertebrate brains have a similar design and process information in similar ways. Brains need various organs for sensing information (for example, the retina), organs for filtering and organizing this information (thalamus) into a form that can be analyzed by an information processor (pallium) to make decisions, possibly with respect to stored memories (hippocampus), and an organ to turn plans into actions (basal ganglia). Underlying this process are organs that keep the rest of the brain ticking over (brainstem) and control the peripheral nervous system, motivating the body to find food, look for sexual partners, and so on (hypothalamus). It is not necessary to understand the details of brain organization or remember the Latin names of different brain regions to appreciate how brains work. Basically, the brain takes in information from its immediate environment. Information is a representation of some aspect of the world, such as the shape and color of a flower, its smell, location, a memory of the last time the

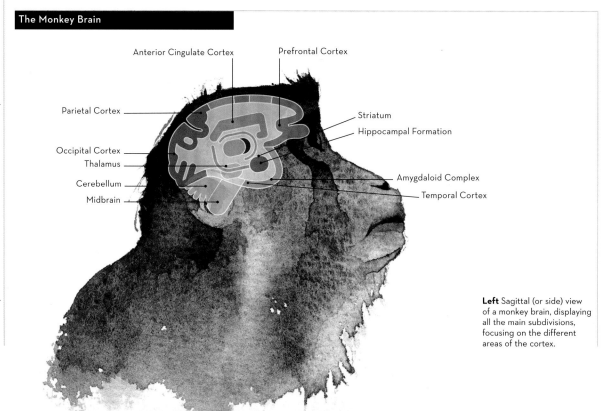

The Monkey Brain

Anterior Cingulate Cortex

Prefrontal Cortex

Parietal Cortex

Striatum

Hippocampal Formation

Occipital Cortex

Thalamus

Cerebellum

Midbrain

Amygdaloid Complex

Temporal Cortex

Left Sagittal (or side) view of a monkey brain, displaying all the main subdivisions, focusing on the different areas of the cortex.

same object was encountered, and the emotion it evoked. This information is interpreted and stored by the brain and used to initiate plans of action. This basic neural architecture is common to all vertebrates.

On the right is a bird (rook) with a schematic of its brain. Different regions are indicated by different colors and given their individual anatomical names. Of most importance to this book are regions of the pallium (hyperpallium, mesopallium, nidopallium, entopallium, archipallium, hippocampal formation), the striatum, and the cerebellum. The rook's brain is about the size of a shelled walnut. On the left is a mammal (monkey) with a brain schematic. Its brain is about the size of a large

plum. It is quite different from the avian brain, most notably in the folded region surrounding the surface of the brain (the neocortex). Avian brains are smooth on the surface, with none of the grooves (sulci) found on chimpanzee or dolphin brains. These sulci represent the folding of the cortex that squeezes more neural tissue into a small space. Imagine trying to fit a sheet of paper into a Ping-Pong ball. The best method is to crumple the paper into a ball and squeeze it inside. Something similar occurs with the mammalian brain where the cortex is crumpled up to fit inside the skull. The avian brain is more like the same sheet of paper being soaked in water and then rolled into a ball to fit inside the ball.

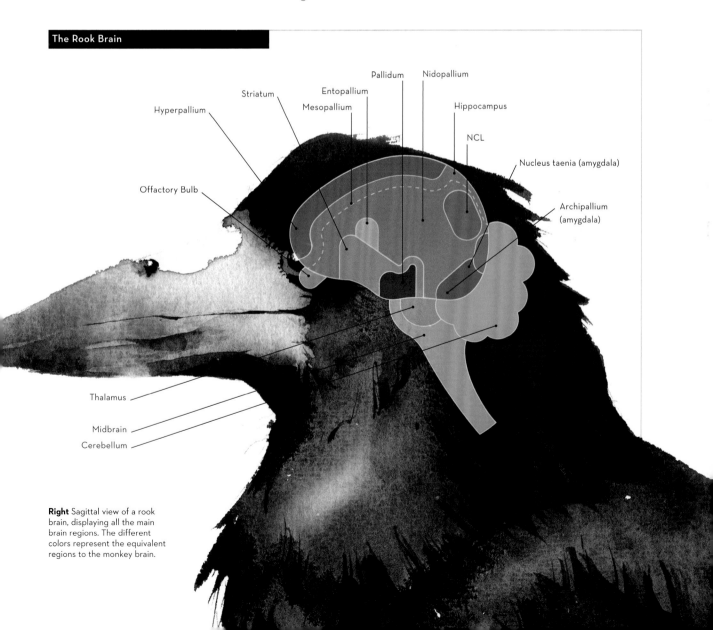

The Rook Brain

Pallidum · Nidopallium · Striatum · Entopallium · Hyperpallium · Mesopallium · Hippocampus · NCL · Nucleus taenia (amygdala) · Offactory Bulb · Archipallium (amygdala) · Thalamus · Midbrain · Cerebellum

Right Sagittal view of a rook brain, displaying all the main brain regions. The different colors represent the equivalent regions to the monkey brain.

BRAINS FOR COGNITION

We need to focus on the neural substrates of thought. The pallium constitutes the largest percentage of the forebrain in both birds and mammals. The question is to what degree is the avian pallium functionally equivalent to the mammalian neocortex? The two contenders are the Wulst (or hyperpallium; a bulge on the top of the brain) and the dorsal ventricular ridge (DVR; consisting the nidopallium, mesopallium, entopallium, and archipallium). When we consider how the pallium has evolved, it has taken a different journey in birds and mammals, but seems to have arrived at the same location in both in terms of function. For now at least, we can assume that a region in the avian pallium is equivalent to the mammalian neocortex.

A SHARED VISION

Aside from shared ancestry, the avian pallium and mammalian neocortex share similar patterns of connectivity. Birds and mammals have two main visual pathways that are similarly connected (see diagram).

The lemnothalamic (thalamofugal) pathway is a direct route for visual information about the location of objects in space to reach the primary visual processing areas of the brain. Information first travels from the retina to the thalamus, which is then projected to the Wulst (birds) or the primary visual cortex (mammals). In parallel, the collothalamic (tectofugal) pathway processes detailed information about the visual features of an object, such as its color, shape, motion, and even its social relevance, such as the identity of a group member. Information passes from the retina to the optic tectum (birds) or the superior colliculus (mammals) to control eye movements, especially in the case of moving objects such as prey. This information is passed on to the thalamus, which directs attention toward the stimulus of interest. Finally, this information reaches the entopallium (birds) or the extrastriate cortex (mammals) for processing visual details. Birds and mammals also share similarities in their auditory (hearing), somatosensory (touch), and motor systems, which suggest similarities in other brain systems, perhaps involved in cognition.

Cat and Pigeon Visual System in Section

Mammals, such as cats (left), and birds, such as pigeons (right) also share similar visual processing pathways. These similar neural pathways allow each species to see in a variety of colors, discriminate between different objects and follow them in motion. These skills allow them to adapt their vision to their different needs.

——— Collothalamic (tectofugal) pathway

——— Lemnothalamic (thalamofugal) pathway

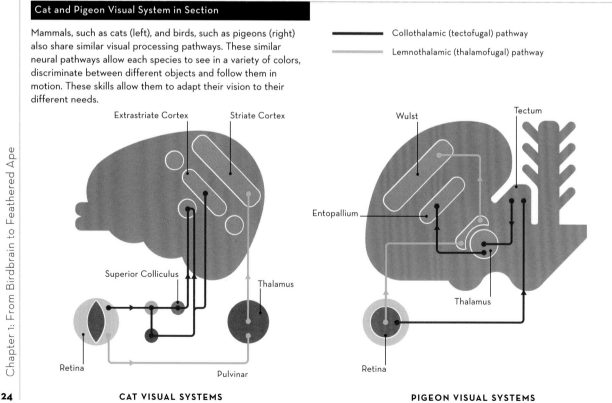

CAT VISUAL SYSTEMS

PIGEON VISUAL SYSTEMS

Below Cats and pigeons are highly dependent on their ability to see. Cats are nocturnal predators that need to see their prey in low light in order to hunt efficiently. Pigeons are diurnal prey that need a wide field of view to see predators, but also search out their primarily grain diet that is small and difficult to see.

Do birds have a prefrontal cortex?

The most critical region of the mammalian brain for intelligence is the prefrontal cortex (PFC), located at the front of the brain. This area has been attributed roles in personality, the theory of mind, self-awareness, problem-solving, and executive functions such as planning, flexibility, and working memory. These are key to what it means to be human.

A CONDUCTOR IN THE BRAIN

Do birds have an equivalent region to the PFC? Studies of behavior, neural connectivity, development, and neurochemistry suggest that the caudolateral part of the nidopallium (NCL), at the rear of the brain, is the avian equivalent to the mammalian PFC. Even the humble pigeon is capable of executive functions traditionally attributed to the prefrontal cortex, namely, working memory, planning, flexible thinking, controlling actions, and attending to objects of interest. These functions are involved in the management of cognition. Imagine that the PFC is the conductor of an orchestra. The conductor can communicate with all other orchestra members, knows what each of them is doing, and can influence what they do next, even though they can function perfectly well individually. The conductor directs the orchestra as a whole, so that it functions to its maximum efficiency, weeding out problems when found and correcting any errors that may occur. This is what the PFC does with respect to the rest of the brain through executive functions.

AN AVIAN PREFRONTAL CORTEX?

Evidence for an avian equivalent to the PFC comes from different types of research. Destroying the NCL causes detrimental effects on working memory and flexible thinking tasks in pigeons. Neurons in the NCL fire most vigorously in the period before a reward is presented in working memory tasks. Connections of the NCL are similar to the primate PFC, linking with multisensory areas, areas responsible for affective (emotional) states and memory (see diagram), as well as the striatum (see diagram). The NCL is substantially larger in some birds (corvids and parrots) than others, suggesting a role in complex cognition. This relationship is also seen when comparing the size of the primate PFC with that of other mammals. Finally, the neurotransmitter dopamine plays an important role in behavior and cognition and floods the PFC with projections from deep within the midbrain. PFC neurons have huge numbers of receptors with an affinity for dopamine. These features are all shared by the NCL.

Despite this evidence, little is known about how the avian NCL is organized. It is not known whether it is subdivided, like the primate PFC, into dorsolateral, orbitofrontal and ventromedial prefrontal regions performing different roles in cognition. As the dorsolateral PFC plays the most prominent role in executive functions, other regions of the avian pallium may represent parallels with other parts of the PFC.

Left Indian mynas were introduced to Australia in the 1860s, and have rapidly colonized urban areas to now been known as one of the worst invasive pests on the continent. Having a large NCL would have provided these birds with the ability to quickly adapt to new surroundings and exploit novel sources of food, thereby making it successful.

Prefrontal Connections

Birds have a brain region called the NCL that is functionally equivalent to the primate prefrontal cortex (PFC), allowing an individual to make appropriate decisions. Both primates and birds can do this by sorting through alternate scenarios in their working memory.

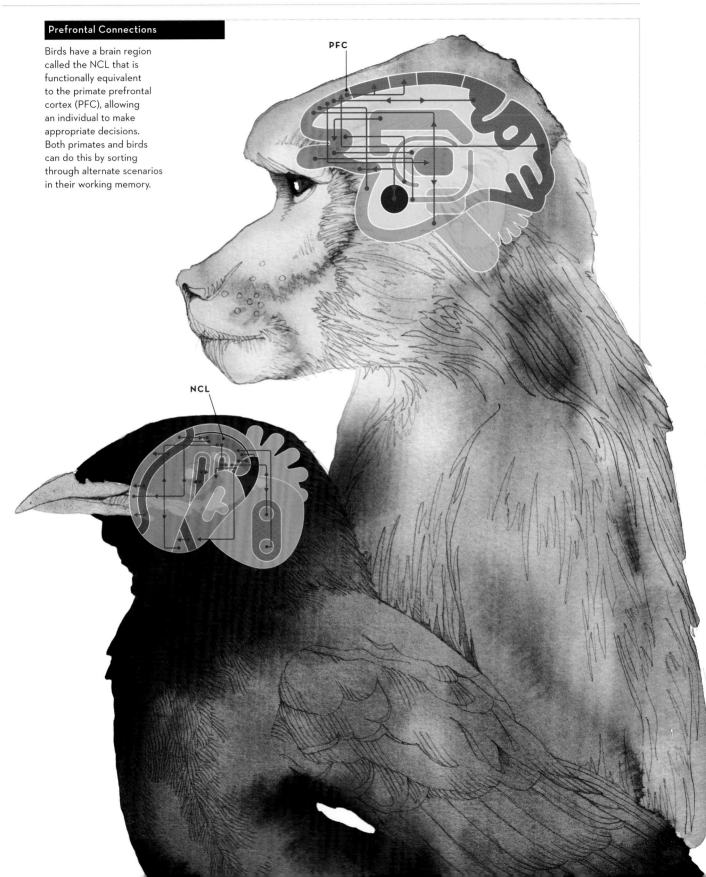

PFC

NCL

How did the avian brain evolve?

Modern birds and mammals are separated by 300 million years of evolution. Their last common relative was a stem amniote similar to a modern day amphibian. All modern families of mammals, reptiles, and birds evolved brains from the basic neural plan in this stem amniote.

LIVING DINOSAURS

Surprisingly, birds not mammals are the most recently evolved animals, although technically, birds are living dinosaurs. In studying how the avian brain evolved, comparisons are made between a typical bird (pigeon) and a typical amphibian (frog), reptile (turtle/lizard), and mammal (rat). We are especially interested in the pallium because that region gave rise to areas involved in perception, learning, and cognition. The forebrain consists of two main parts: the pallium and subpallium. The pallium gave rise to the cortex, hippocampus, and amygdala (involved in sensory processing, thinking, memory, spatial navigation, and emotion), whereas the subpallium gave rise to the basal ganglia (involved in learning habits and species-typical behaviors such as sex, feeding, and parenting). We will focus on the pallium.

The basic amniote brain plan is displayed in the diagram. There are three parts to the pallium (dorsal [top], medial [middle], and lateral [side]), which sit atop the dorsal ventricular ridge (DVR). The pallium sits above the striatum in the subpallium, and the septum sits between pallium and subpallium.

THE IMPORTANCE OF PALLIAL EXPANSION

During evolution, two different routes transformed the pallium in reptiles and birds on the one hand and mammals on the other. In mammals, a dorsalization of the pallium occurred in which the dorsal regions expanded in size to form the neocortex. The ventral pallium (DVR) remained small with the medial pallium forming the hippocampus, the lateral pallium forming the olfactory cortex, and the DVR forming the amygdala. In reptiles and birds, the reverse was true, with a ventralization of the pallium. In reptiles, the medial, dorsal, and lateral

pallial regions remained relatively small, whereas the DVR expanded. Similarly, in birds, the ventral pallium expanded into a massive DVR and although the dorsal pallium evolved into a large Wulst (especially in birds of prey), other areas of the dorsal pallium remained relatively small.

Despite differences in how the pallium evolved, both birds and mammals rely on this structure for their cognitive abilities. How is this possible? Scenario 1 suggests that the avian pallium and mammalian neocortex took different evolutionary pathways, with vastly different structures, but which ended up solving similar problems. Scenario 2 suggests that the avian pallium and mammalian neocortex are homologous, basically the same structures, thus explaining why they can solve similar problems. It is not clear which of these evolutionary scenarios is best explained by the data. Fundamentally, avian brains are nucleated with no laminated cortex, whereas mammalian brains have a six-layered cortex, and thus are physically different, though, as we have seen, they also share many common features in the way they are wired up.

Above Birds share features with both lizards and mammals, and this is reflected in the evolved structure of their brains.

Right Despite looking and behaving very differently, all reptiles, birds, and mammals share the basic neural organization of our common ancestor, very like a modern-day amphibian such as this toad.

Evolution of the Vertebrate Pallium

These slices through the brain show the left hemisphere looking straight on from the front. They represent a schematic organization through the forebrain of a typical stem amniote, mammal, reptile, and bird, to show similarities across brain regions in these animal groups.

- Dorsal pallium
- Medial pallium
- Lateral pallium
- Dorsal ventricular ridge
- Septum
- Subpallium
- Ventricle

STEM AMNIOTE   **MAMMAL**   **REPTILE**   **BIRD**

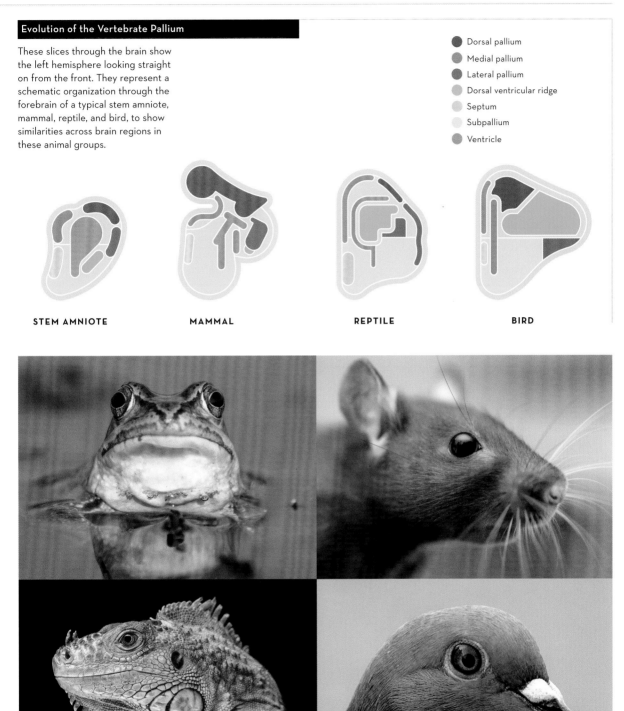

Computers, cakes, and cubes

How could such different structures—the avian brain and the mammalian brain—function in such similar ways? Two analogies might help.

COMPUTERS

Let us think of the avian brain as like an Apple Mac computer. It has a power source, a microprocessor, a keyboard and a mouse to input information, and a monitor to display that information. It runs software using a specific programming language written specifically for its processing system. Now think of the mammalian brain as like an IBM PC. It also has a microprocessor and peripherals, but the way that information is processed inside is very different in the Mac compared with the PC. You can input information into the Mac and get the same outputs as the PC, but that information will be interpreted in different ways inside the Mac because its programs are written to handle data differently. Mac programs are written in one language and PC programs written in another. However, once each microprocessor has made its calculations, the output may look very similar, especially now, since PCs have reverted to using a graphical interface.

Something similar may occur in the way avian and mammalian brains work. The inputs are the same (sensory information) and the outputs are very similar (behavioral responses, cognitive operations, and so on)

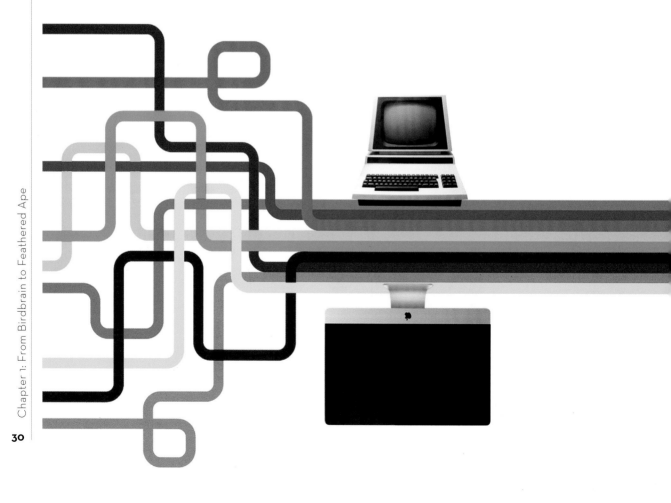

and yet what is going on inside the information processor (brain) is quite different. This variation is driven largely by differences in the hardware of the brain (as we know little about the software).

CAKES

With regard to the hardware, another analogy suggests that the mammalian cortex resembles a sponge cake with six layers, whereas the avian pallium resembles a fruitcake, with no real layers, only clumps of fruit (nuclei) found throughout the cake (brain). The same ingredients have gone into making the cakes (eggs, flour, fruit, sugar, and butter equal neurons, glial cells, and so on), but different recipes and baking (evolution) led to radically different results.

CUBES

To visualize this in 3D, compare two cubes. In the mammalian brain cube (upper), the majority of neurons are found around the surface in a series of six layers. The rest of the brain consists of the connections between neurons in these layers (and with the underlying subcortical structures). The neurons in the cortical layer near the surface are the gray matter, whereas the connections throughout the brain are the white matter. The avian brain cube (lower) does not have a substantial amount of white matter because neurons are found throughout the brain, as there is no layered cortex. Nuclei with similar functions group together with short connections between them; there are few long connections in the avian brain. What this means for how information is processed in these two types of brains is not yet known.

Left Although Macintosh computers and PCs process information differently (represented by the different colored lines), the resultant outputs, playing a game, calculating a series of numbers or writing a document, all converge on a similar solution. This is similar to the processing of mammalian and avian brains.

Mammalian and Avian Brain Cubes

LAMINATED CELL CLUSTERS
Gray matter (layered) at edge
White matter in center
Gray matter (nuclear) in between

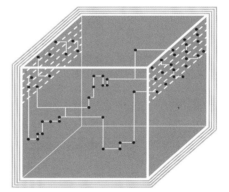

MAMMALIAN BRAIN ORGANIZATION

NUCLEATED CELL CLUSTERS
Gray matter throughout brain

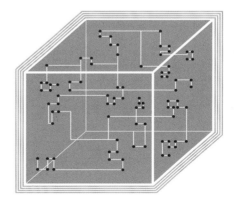

AVIAN BRAIN ORGANIZATION

Does brain size matter?

Birds tend to be small. Even the largest bird, the ostrich, is a mere two percent the size of the largest land mammal, the African elephant.

SMALL BODIES, BIG BRAINS

Birds are small and light because most birds can fly and, in order to fly efficiently using the least amount of energy, birds' bones are hollow, very light, and yet strong. All parts of birds are small and light, including their brains. However, they compensate for this reduction in mass by using a few nifty tricks, such as generating new neurons when they need them. Birds are not all equal. Despite significant differences in body size, from the smallest, the bee hummingbird (1/5 oz/6 g), to the largest, the ostrich (270 lb/123 kg), they differ not only in absolute brain size (actual weight of the brain) but also in relative brain size (size of the brain relative to body size). Bigger animals tend to have larger brains, because brains like all other organs scale up with body size. Yet birds have to minimize their overall body size. Whales and dolphins (cetaceans) do not have this problem because salt water supports their bodies, so size is not a constraint. Hence most whales are many times larger than the largest land mammal and grow a disproportionately large brain. This has led some to suggest that dolphins have received an overly generous press with regard to their intelligence because in reality their brains function not for intelligence but in regulating body heat in the cool aquatic environment. But this argument seems unlikely, because not all aquatic animals have relatively large brains and not all cetaceans are blessed with complex cognition (largely because their lives are not filled with challenges requiring superior brain power).

PIGEONS ARE NOT CROWS

Pigeons are not the mindless automatons envisaged by the public. Although it is true that they are not capable of solving complex problems—communicating using symbols, predicting another's actions, or imagining the future—they are exceptionally gifted at discriminating between different objects. This is likely because their day-to-day foraging requires them to locate small items of food from an intricate textural background and to distinguish foods from non-foods. Pigeons are also exceptional learners with impressive memories. Although not the fastest in the class, they pick up information relatively quickly and retain huge amounts over long periods—for example they are able to remember hundreds of pictures more than two years later. Despite these skills, pigeons do not approach crows and parrots in terms of their cognitive skills. Why might this be?

Birds differ in absolute and relative brain size and also in their intelligence. Some birds of a similar size, say crows and pigeons, differ dramatically in overall brain size. Crows have a brain twice the size of a pigeon's brain (see figure). However, brains do more than just thinking. A significant amount of the brain controls bodily functions, such as heart rate, muscle tone, and breathing, as well as sensing the world and instinctual behaviors that rely little on cognition. Regions in the pallium, such as the mesopallium and nidopallium, are dedicated to learning and cognition and their relative size influences a bird's intelligence. The nidopallium varies greatly in size across species. For example, the nidopallium in rooks is three times larger than in pigeons, even though pigeons and rooks are roughly the same body size. Rooks also pack three times as many neurons into each cubic millimeter of the nidopallium compared with pigeons (to the tune of 180,000 as opposed to 60,000). With this larger nidopallium, rooks achieve more flexible, intelligent behavior than has been demonstrated in pigeons.

We don't know how these differences in region size and neuronal density equate to differences in intelligence, aside from the fact that bigger brains equal more neurons, which means a more efficient computer. But size is no longer relevant. Indeed, the computer in my smartphone is more powerful than my old PC laptop.

Bee Hummingbird and Ostrich

Despite vast differences in their body size, the proportion of the body taken up by the brain is significantly larger in the bee hummingbird than in the ostrich. The hummingbird has a brain much larger than expected for its tiny size.

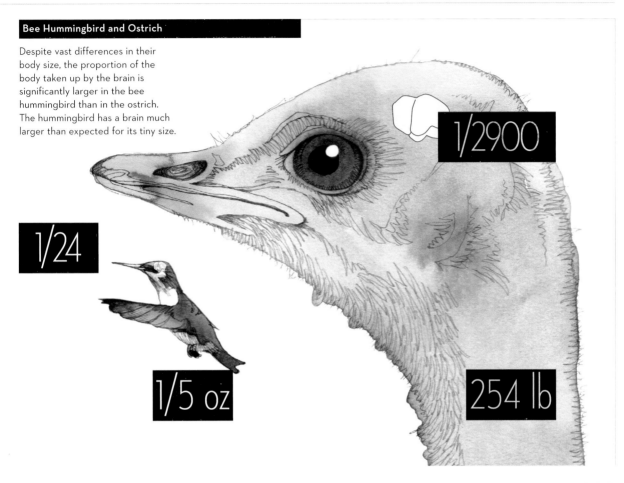

1/2900

1/24

1/5 oz

254 lb

Brain Comparison

A jungle crow brain (left) compared with a pigeon brain (right). The crow brain is about twice as large as the pigeon brain, despite the birds being about the same body size. Most of the additional brain is focused in the pallium.

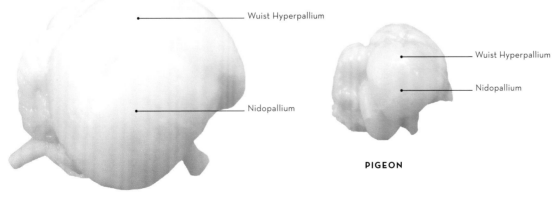

Wuist Hyperpallium

Nidopallium

Wuist Hyperpallium

Nidopallium

PIGEON

JUNGLE CROW

Were dinosaurs smart?

Can we say anything about a species' intelligence from their brain size? We know from living species that absolute brain size is an inadequate indicator of intelligence, because there are animals with very large brains that could be classified as dumb and animals with very small brains that could be classified as smart.

THE CLEVER CLUB

Blue whales have the biggest brains ever to have evolved, but these are tiny compared to their bodies. Contrast this with killer whales, which have very large brains relative to their body size, although these are absolutely smaller than those of blue whales. Differences in the challenges requiring a cognitive solution faced by these two species may provide a clue as to why their brains are so different. Blue whales are solitary and feed on plankton collected by opening their huge mouths. By contrast, killer whales are very social, hunting seals in cooperative pods, and possibly demonstrating culture, so the need for a smarter brain is greater in killer whales than blue whales. Other animals, such as elephants, chimpanzees, parrots, crows, dolphins, and humans all have brains larger than predicted from their body size, and all face similar challenges in their environment. For this reason, we could classify them all as members of the Clever Club.

WHAT ABOUT DINOSAURS?

Some dinosaurs had feathers, could fly, walked upright, and laid eggs in nests. Birds are the closest living relatives to dinosaurs. Can we make any inferences about dinosaur intelligence from birds, such as their brain size? We have already noted that we cannot directly infer an animal's intelligence from the size of their brain because the organ controls many functions aside from cognition. Even if we can measure the relevant brain region—the pallium—it is still conjecture for us to say anything about an extinct animal's behavior and cognitive abilities from brain size alone.

Although we have data on dinosaur brain size, it is not perfect, as calculations of body size have to be estimated from partial skeletal remains and brain size is calculated from endocasts made from inside the skull. However, approximations of relative brain size can be made. All dinosaurs had tiny brains for their body size,

with the massive sauropods (such as brachiosaurus) having the smallest brains. We may predict that the most recently evolved dinosaurs, such as velociraptors—those most closely related to living birds—should have had the largest brains and so would have been the most intelligent. In the movie *Jurassic Park* velociraptors were shown opening doors and solving problems in a similar way to primates. This attractive idea is not backed up by any data, as velociraptors were about the size of overlarge turkeys and although their brains were large compared with other dinosaurs, they were tiny compared with those of modern birds and mammals. Indeed, their brains were smaller than ratites (emus and ostriches), creatures not renowned for their intelligence. Although velociraptors must have faced some of the same challenges as modern birds, they probably faced those challenges using brute strength rather than their wits. So, it's unlikely that any dinosaur would have been capable of the intellectual feats displayed in the movie.

SPECIES	BODY SIZE	BRAIN SIZE
Tyrannosaurus rex	13,670 lb (6,200 kg)	7 oz (200 g)
Velociraptor*	22 lb (10 kg)	$1/8$ oz (3 g)
Archaeopteryx	$7/8$ lb (0.4 kg)	$1/16$ oz (1.5 g)
Ostrich	270 lb (123 kg)	$1^1/2$ oz (42 g)
Raven	$5/8$ lb (1.2 kg)	$1/2$ oz (15 g)
Macaw	3 lb (1.4 kg)	$7/8$ oz (24 g)
Elephant	5,620 lb (2,550 kg)	160 oz (4,500 g)
Dolphin	395 lb (180 kg)	60 oz (1,650 g)
Blue whale	397,000 lb (180,000 kg)	315 oz (9,000 g)
Chimpanzee	115 lb (52 kg)	15 oz (430 g)
Human	145 lb (65 kg)	50 oz (1,400 g)

estimated from volume of endocast of close relative, Tsaagan.

Brain and Body Size Across Birds and Mammals

The blue whale has the largest brain of any creature, but it also has an enormous body to manage. Tyrannosaurus rex, like all dinosaurs, had a relatively tiny brain responsible for running a big body. Humans have a proportionally large brain in relation to the size of their bodies, the largest of any primate.

1/20000

1/31000

1/46

THEORIES OF BRAIN AND INTELLIGENCE

Due to the energy required for flight, birds evolved brains able to maximize efficiency in processing information using the fewest neurons or the shortest connections. Birds with brains larger than predicted for their body size, such as crows and parrots, must put them to good use, as they are expensive, using 20 percent of the body's energy requirements.

Different selection pressures are proposed to explain the expansion of some birds' brains and their subsequent intelligence. Did they grow bigger to cope with living in larger social groups or in response to ecological challenges, such as locating distributed food? Or did brains get bigger to allow birds to solve novel problems and exploit new habitats?

SOCIAL SMARTS

The most popular theory is the social intelligence hypothesis, which posits that social animals need a larger brain to process information about group members—who are their friends and who their enemies; the history of their relationships—and to give them the ability to deceive one another and predict the intentions of others. Big brains could store this information and make more detailed computations. In primates, the more social species tend to have larger brains. However, a similar relationship is not found for birds. Although some birds form huge flocks, these do not remain stable, as individuals form pairs during the breeding season, then separate after mating or raising offspring. There is no need to remember individuals and their social histories. What is more important is maintaining a long-term, selective relationship with another individual (pair-bond). The relationship intelligence hypothesis suggests that skills essential for keeping a pair "in tune" with one another, through sharing, copying, and collaboration in defending a nest, raising, and teaching offspring, require a larger brain than living in a large flock.

FOOD FOR THOUGHT

Alternative theories focus on challenges of the physical environment, especially finding and processing food. The spatio-temporal mapping hypothesis suggests that animals who eat seasonal fruits, ripening at different times and locations, require a brain that remembers where and when trees produce ripe fruit. There is evidence for this hypothesis from fruit-eating monkeys with brains twice the size of leaf-eating monkeys, but no direct evidence from birds. However, birds such as scrub jays, possessing what-where-when memory for the type, location, and timing of hidden food (see Chapter 2), tend to have relatively large brains, equivalent in size to those of our earliest ancestors, the Australopithecines.

TECHNICAL SAVVY

Other theories focus on processing food, namely the extractive foraging and tool-use hypotheses. The former suggests a bigger brain provides the knowhow for extracting the contents of encased foods, such as nuts, shells, and hard-skinned fruits. The latter suggests that species that use tools to extract these foods have bigger brains. No one has studied the former in birds, but there is evidence for the latter. Birds that use true tools (that is, detached objects that function to achieve a goal; see Chapter 5) have larger relative brains than those who do not use tools or only proto-tools. True tool use would be an Egyptian vulture using a rock to break open an ostrich egg, whereas proto-tool use would be a thrush bashing open a snail on a pavement, though both would be examples of extractive foraging. One final theory is the innovation hypothesis, relating bigger brains to the ability to produce novel behaviors or solve novel problems. Innovation permits an animal to invade new environments, exploit new resources when others are in short supply, and cope with change, such as climatic variability.

Although some birds have large brains, there is no one convincing theory to explain why. Large-brained birds often form lifelong pair-bonds, but within large flocks, with a varied diet including encased foods that

Above An Egyptian vulture dropping a rock onto an ostrich egg to break it open and eat the tasty contents. This is an example of tool use, as the vulture drops a rock (tool) onto the egg. Dropping the egg onto a rock would not be an example of tool use.

frequently require tools to extract, as well as new resources and environments needing the development of innovations to exploit. Behavioral flexibility—updating behavior to address changing circumstances and adapting to new challenges—may underlie all these traits. This idea, which can be tested with experiments, may provide a better explanation for the evolution of intelligence across birds and mammals than currently accepted theories.

Right Parrots, such as these scarlet macaws, are one of the most social animals, hanging out in large colorful flocks. They keep together by using a sophisticated system of calls, with some calls used like human names.

FEATHERED APES?

As we have seen throughout this chapter, our understanding of the evolution of the avian brain has changed dramatically over the last one hundred years. Our understanding of bird cognition has also changed just as dramatically.

THE AVIAN POSTER BOY

Up until the 1970s, pigeons were the avian poster boys, with our understanding of avian intelligence restricted to what was known about pigeon intelligence. Pigeons were thought to be excellent learners, with skills in discriminating between different stimuli that are often superior to our own. However, pigeons were also perceived as brainless simpletons pecking out a meager existence in our town squares and high streets. Thankfully, careful studies of not only pigeons but a wide variety of birds, including parrots, crows, and other songbirds, have revealed that some species should be reassessed and could even be considered feathered apes. This term was introduced in 2004 to reflect the fact that some members of the corvid family demonstrate cognitive skills that are on a par with those of the great apes.

Crows and apes are separated by more than 300 million years of evolution. Almost none of the relatives that have emerged in the interim—reptiles, most other birds, and even mammals—possess the same cognitive skills, suggesting that any similarities between crows and apes have arisen through convergent evolution. The rest of this book will reveal some of the intellectual feats of feathered apes, and how these creatures compare with great apes and even humans.

A SMARTER BRAIN

Why feathered apes? The term suggests birds with the same minds as apes, but within the limitations of a differently structured brain. Are the minds of crows and apes the same, and how would we know? The relative brain sizes of these two groups are similar, so in terms of their overall body size, crows and apes have brains that are larger than predicted for their body size. An ability that may set members of the Clever Club apart is a flexible form of generalized thinking that can be applied outside of the context in which it evolved. Some non-tool-using birds understand the processes underlying how tools work (see Chapter 5).

Above A New Caledonian crow has made a stick tool and uses it to probe for grubs hidden inside a tree trunk. After a period of probing with a tool, the agitated grub grips hold of the end of the stick tool and the crow can pull it out and eat its tasty treat.

Unfortunately, cognitive studies comparing vastly different species are still in their infancy and the tools that will allow us to compare creatures, without disadvantaging species based on differences in their perceptual skills or their ability to manipulate objects, are still being devised. It is relatively safe to compare apes and corvids because both have well-developed visual skills (see Chapter 3) and can manipulate a variety of objects. Indeed, crows are not as disadvantaged in this area as might be expected with only a beak and no hands. My wife and I experienced this directly when a hand-reared young rook quickly unscrewed a nut and bolt, tightened by a human using a wrench!

The aim is not to compare taxa in a competitive manner—this is not an intellectual animal Olympics to see who tops the podium—but rather to discover the form that learning and cognition takes through careful experiments to see how species are similar, how they differ, and how these differences may be attributed to differences in brain structure.

Right Most of what we know about the avian mind and brain is due to studies of pigeons, but unfortunately they are not considered the smartest of animals, never mind birds.

2 WHERE DID I HIDE THAT WORM?

How birds navigate

Animals that travel need the ability to navigate. For birds, this can be a huge problem, as the distances they can traverse are greater than for any other creature, potentially spanning the entire globe. Birds also travel through some of the most complex three-dimensional environments on Earth, especially compared to those traveled by ground-dwelling mammals.

FINDING YOUR WAY AROUND

Finding its way around is one of the first skills a young bird develops when it leaves the nest. For newborn chicks of precocial species, such as geese and chickens, this is a problem faced straight after hatching, as they have to fend for themselves, only finding food by following their mother. Young chicks do not need a spatial ability; rather, they become imprinted onto the first moving object they see upon entering the world, usually their mother. Once the chicks have left their mother's protection, they are on their own and have to find their own way in the world. For altricial species, such as birds of prey, parrots, or crows, where the chicks are fed and looked after by one or both parents, the chicks remain in the nest much longer until fledging, when they can start to find their own food.

SPATIAL PROBLEMS

Most of life's problems require navigation. You have to find your way back to your nest, especially if you have chicks to feed and protect. If you have found a good place to forage, it is better to return the next day rather than begin the search anew. Spatial problems range from short to long distances, and can dominate some birds' entire lives, influencing the design of their brains and the way they use them. Some examples of the spatial problems faced by birds are: migrating each year or season to return to breeding grounds that provide enough food for new offspring; pigeons finding their way back to their home loft; birds hiding large numbers of seeds and finding them months later when food becomes scarce, or hiding food that perishes and is prone to pilfering; birds surveying other birds' nests (brood parasites, such as cuckoos), remembering the

location of potential targets for laying; birds feeding new chicks needing to remember the location of their nests so that they can return to the correct one; and hummingbirds foraging on flowers remembering which flowers they just ate from and when they will provide nectar again.

SWIFT SKILLS

In most of these cases, finding the way is a matter of life and death. On a trip to Iguazu Falls in Brazil, my wife and I were impressed by swifts foraging on flying insects at dusk around the falls. Swifts roosted behind the falls, but they continued feeding until the light levels were so low that their trip back to roost became very dangerous. The swifts had to weigh up the odds of finding their roosting spot in the dark and getting enough food to eat to sustain them through the cold night sleeping behind a raging waterfall. It was amazing to see the birds plunge behind the falls, even when almost dark, successfully finding the right location without bashing their heads in! They must have known where to go, but whether they formed a picture of the environment behind the waterfall and the location of their roost; or they flew in the general direction of their roosting spot and found it once behind the falls; or they didn't have a specific roosting spot but just fought it out once they got there, is still unknown. However they do it, these swifts have evolved an impressive spatial ability.

In this chapter, we will investigate how birds navigate; how they use their often impressive memories to find their way around, to find food (including food they hid themselves), and even to remember the specific time and place they encountered or hid a specific type of food—something resembling human episodic memory. However, we will start from the opposite end of spatial memory, the sometimes unbelievable global journeys or migrations performed annually by birds, some only as big as a human thumb.

Left The Arctic tern has one of the longest migratory flights of any bird, sometimes travelling up to 40,000km across the entire length of the world from northern Arctic regions to the southern Antarctic to breed in food-rich waters and make the return trip when their young are ready to fly.

A life on the wing

Imagine you are a small songbird living in a North American suburb, such as an American robin. As temperatures drop down to freezing as winter approaches, you suddenly up sticks and fly south to the most southern reaches of the continental United States and beyond, eventually reaching all the way to food-rich breeding grounds in Guatemala. Here you will stay until the following spring when your offspring are raised and temperatures back home are warm enough for your primary food, insects and fruit, to become plentiful again.

ITCHY WINGS

For migratory birds, the decision to travel such long distances each year is driven not by their psychology—they do not think that it is time to leave—but by changes in hormone levels related to changing day length, reductions in temperature, and decreases in the amount of food available. These endocrine changes ultimately cause physiological and behavioral changes, and initiate what is termed migratory restlessness or the urge to travel.

The annual trip of the American robin is nothing compared to some birds. Perhaps the most striking case is the Arctic tern flying from feeding grounds in the Arctic Circle, along either the east or west coast of North and South America or down the coast of Africa to new feeding grounds in Antarctica, before making the return journey once its chicks have fledged. This amounts to a round trip each year of between 45,000 and 55,000 miles (70,000 and 90,000 km), with most of it spent on the wing. As can be seen in this world map, most of the birds performing such long migrations do so initially from north to south, then back to the north, following changing environmental and climatic conditions, and their subsequent effects on food supplies and opportunities for breeding.

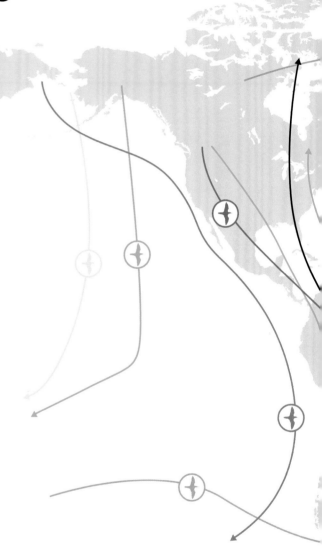

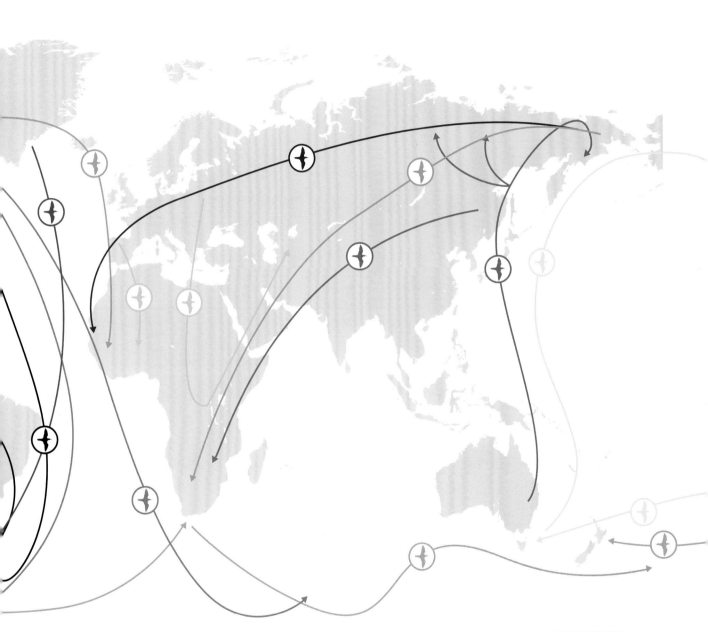

Migration Routes

Avian long-distance migratory routes. Each colored line refers to a different species, with the approximate length of the route noted.

— Ruff: 18,600 miles (30,000 km)

— Far Eastern curlew: 3,700 miles (6,000 km)

— European bee-eater: 6,500 miles (10,500 km)

— Amur falcon: 13,700 miles (22,000 km)

— Cape petrel: N/A

— Arctic tern: 23,600 miles (38,000 km)

— Manx shearwater: 6,800 miles (11,000 km)

— Northern wheatear: 18,600 miles (30,000 km)

— Short-tailed shearwater: 6,200 miles (10,000 km)

— Bar-tailed godwit: 7,300 miles (11,700 km)

— American golden plover: 15,500 miles (25,000 km)

— Bobolink: 11,800 miles (19,000 km)

— Swainson's hawk: 14,300 miles (23,000 km)

As the crow flies

How do birds perform the seemingly impossible task of finding their way across vast distances, often with few features to guide their passage, and with no technology to help them? Migratory birds have evolved a natural suite of navigational tools—the equivalent of compasses, satellite navigation, and GPS.

Below Blackcaps migrate either southwest from Germany, or southeast from Hungary, eventually wintering in breeding grounds in West or East Africa. Their offspring will also follow these migration routes. Chicks raised by the parents of a different population, such as southwest chicks being raised by southeast parents, will lead to chicks taking a migration route directly south.

Without our advanced technology, we would have trouble finding our way across 5,500 miles (9,000 km) of featureless ocean, and even with it we'd eventually run out of battery or lose satellite signal. Indeed, until John Harrison invented the marine chronometer in 1735, sailors were unable to make safe sea voyages across long distances without relying on calculations of lunar distance that are prone to measurement problems on board an unstable ship.

MIGRATORY GENES

Birds migrating such long distances or even the comparatively shorter distances of homing pigeons employ a number of different mechanisms to help them find their way, which are used at different stages of the

Below Pigeons use different sources of information in their environment to find their way around, from the position of the sun in the sky, and the Earth's magnetic fields, to large landmarks, such as mountain ranges and forests, and smaller landmarks, such as roads, rivers and individual buildings. Some cues are more important when setting off on a journey, and some when close to home.

trip. After getting the impetus to travel, how does a bird know which way to go? Surprisingly, it appears to be under genetic control, with a locus on the ADCYAP1 gene. In a classic study, two populations of blackcaps migrated in different directions: one northern European population flew southwest to winter initially in southern Spain, before moving on to equatorial Africa; whereas the other eastern European population migrated southeast toward Turkey and eventually East Africa. When birds from each population were allowed to breed, their offspring contained genes from both populations and flew due south between the migratory routes of their parents.

SIGNS OF HOME

At this stage, we know more about the mechanisms that pigeons use to home, than other birds use to migrate, so we will focus on pigeons. Once a pigeon has set off from home, it encodes a map of its local environment, based on the arrangement of landmarks around, and leading away, from their home roost.

By constructing such a map, the bird guides itself back home on the return trip. It constructs this map initially from large landmarks, such as buildings and roads (but also around sources of food and water), and is stored in the hippocampus (see page 52-53). Once out of sight of home, the bird shifts from using a cognitive mapping system to a physiological compass-based system for the longer stages of the journey. It would be impossible for birds traveling thousands of miles to store a map of all the landmarks along their route. Therefore, depending on conditions, birds employ a sun, celestial, or magnetic compass to guide them on their way.

SUN CLOCKS

Migratory birds use the sun to orient themselves, via their internal clock, and the actual position of the sun in the sky. Because the sun does not remain fixed, it can only be used as a navigational guide in conjunction with a mechanism for telling the time. Birds have three internal (circadian) clocks—in the retina, the pineal gland and the hypothalamus in the brain—which can process basic information about the time of day using cycles of light levels; the sun is at a specific position at a specific time of day, as gauged by the available light. To determine the role of the sun in navigation, birds are clock-shifted, meaning they are confined to a light-sealed room for a number of days, but provided with artificial light periods different from normal daylight. For example, the bird may receive hours of light stimulation when it is actually dark outside. Birds are then displaced, released and their ability to orient home is tracked. Clock-shifted birds will head in the wrong direction based on the amount of time by which their internal clock has been shifted. So, if the bird was clock-shifted by six hours, it would head 90° in the wrong direction.

NAVIGATING BY STARS

Not all birds migrate during the day, and some continue to fly day and night. How do they orient themselves without the sun?

Nocturnal birds use a celestial (star) map as a source of directional information. Experiments with migratory birds in planetariums have found that they learn celestial maps based on the position of a certain number of the major constellations, and their position relative to the poles. When faced with a simulation of the northern hemisphere sky in the spring, birds will orient north, and will orient south when presented with the northern sky in the autumn. If the birds are presented with the southern hemisphere sky, then they will do the opposite. Birds don't seem to orient using the obvious North Star, the only fixed point in the night sky, probably because at some times of the year it is not visible.

MAGNETIC MIRACLES

Celestial objects, such as the sun and stars, are conspicuous but not fixed objects, so other mechanisms, such as an internal clock, are required to make use of them for navigation. By contrast, geomagnetic fields provide a relatively fixed map of horizontal space across the surface of the Earth, and are probably the reason why so many birds are capable of long-distance migrations. The suggestion that birds use magnetic fields to orient themselves has been around for a long time, but was not confirmed until studies in the 1970s found that attaching a magnet to a pigeon caused it problems in finding its way home on a cloudy day, whereas a brass bar did not.

Two mechanisms have been proposed for how birds use magnetic fields. First, magnetoreceptors called cryptochromes have been discovered in the birds' retina (primarily the right eye). This is equivalent to a chemical compass that is responsive to changes in magnetic directions. A specialized group of neurons in Cluster N, an area of the dorsal hyperpallium next to the Wulst, may process information from cryptochromes. Cluster N neurons become active in response to magnetic cues, and Cluster N lesions eliminate a bird's ability to orient using magnetic fields, whereas their use of the sun or celestial compass is unaffected. Second, iron oxide or magnetite (tiny magnets) has been found in the nasal cavity and skin of the upper beak of pigeons. Magnetite aligns with the magnetic field, providing information as to the bird's position, forming part of its navigational toolbox. Strong magnetic pulses introduced to migratory silvereyes caused them to change direction by 90°. The opthalmic branch of the trigeminal nerve innervates the magnetite-rich regions in the beak and nasal cavity, but it is still unclear how changes in the magnetite are perceived and translated into information that can be used by the brain.

Pigeon Magnetic Sense

Pigeons use a variety of different internal (hippocampus, mechanoreceptors, magnetite) and external (sun, celestial, and magnetic compasses) mechanisms to help them fly home to roost.

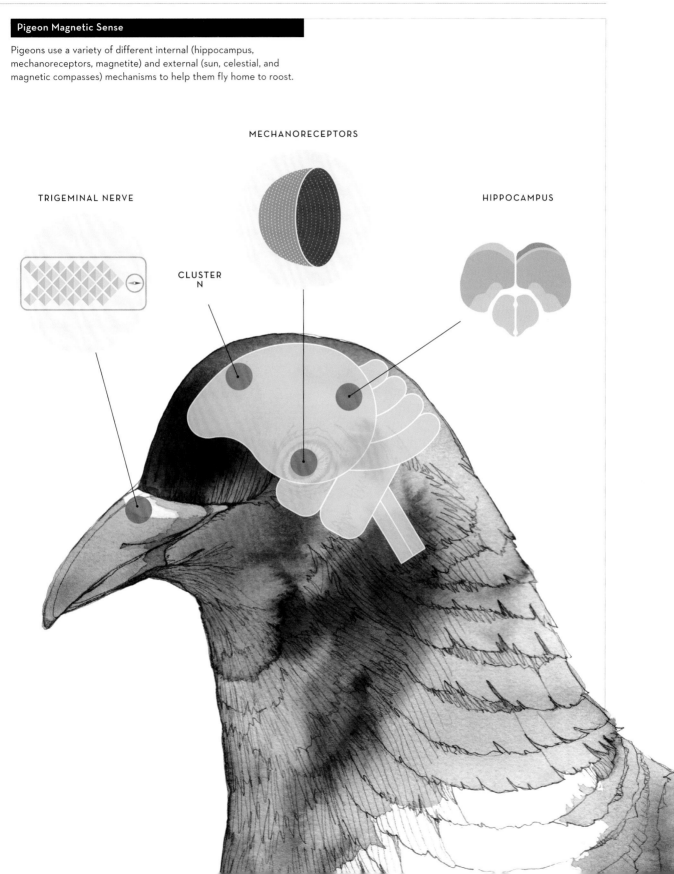

MECHANORECEPTORS

TRIGEMINAL NERVE

HIPPOCAMPUS

CLUSTER
N

REMEMBRANCE OF SPACES PAST

Most birds don't migrate, but that doesn't mean that demands on their spatial memory are less than for migratory birds. Indeed, migration is based on cues that do not necessarily require memory. Those behaviors requiring an ability to navigate on a smaller scale, such as finding new sources of food and then returning to a home roost, or finding hidden food caches, are more likely to be dependent on spatial memory.

GEESE IN A CART

Animals use different forms of spatial awareness depending on the cues available to them. All mobile animals, including invertebrates, are capable of finding their way home without any cues using a process called dead reckoning or path integration. Have you ever been lost in an unfamiliar town where all the buildings look the same but eventually found the way back to your car without knowing how? This is dead reckoning in action. Animals, such as ants foraging in a desert, take a winding path before they find food, but seem to take a more direct route home. There are no landmarks to guide them, rather they keep track of every change in direction and the distance traveled until they reach their goal. They then integrate this information into the best vector directing them to their starting point (all done without thinking). Geese were transported in an uncovered cart from their home to a novel location A. Then, their cart was covered and they were taken to a new location B. When released, they walked in the direction that home would have been from location A, suggesting they had formed a vector of the route to A which they could see in the uncovered cart, but ignored additional information from the second route that they couldn't see. Path integration in birds may, therefore, be dependent on vision.

GUIDING THE WAY

Many animals when encountering an environment for the first time memorize key features and their relationships to one another, constructing a navigational map. These features are separated into beacons, conspicuous features of a goal that are close (proximal) to it, and landmarks, larger objects located further (distal) from the goal which are used to form a picture of the goal's location. Nestling chirps are an example of a beacon to a predator trying to find its next meal or to parents trying to bring back food for their hungry offspring. A Eurasian jay hides food equidistantly between a tall fir tree, a large group of rocks, and a hedge, and the arrangement of these landmarks helps the jay accurately locate its hidden stash.

There are two theories as to how animals use landmarks to find their goals. One landmark is not sufficient to find a goal, because it only provides information about the distance between the landmark and the goal, not the goal's direction. Given two or more landmarks, their relationship to a goal can be used to pinpoint the goal more accurately. In the vector sum model, the angle between two landmarks provides enough information to locate a goal, but each landmark is used in turn. Pigeons were trained to search for food relative to two landmarks that were equidistant from food. One of the landmarks was then shifted to the left. The pigeons computed an average distance in relation to the new positioning based on the landmark that had moved and the fixed landmark, and searched there for food. This is akin to path integration but with landmarks. In the multiple bearings model, a more cognitive method is suggested, as the goal is found in relation to a number of landmarks and the bearings taken in relation to them. Rather than having to compute a vector for each landmark in turn, specific bearings taken between multiple landmarks as a whole can locate a goal. Adding more landmarks ultimately increases search accuracy. Clark's nutcrackers, for example, appear to use multiple bearings when using landmarks to locate their hidden caches.

Right Clark's nutcrackers are corvids that live in extremely tough environments, at a high elevation with a harsh climate. As such, they have to plan for the time when there may be little food to eat during the winter months, so they hide lots of foods, such as seeds they can eat throughout the year when times are tough.

Wild memories

Rufous hummingbirds in the Rocky Mountains have a difficult spatial problem affecting their survival. They feed on flower nectar, but each flower is a limited resource that only replenishes itself after a certain period of time.

FORAGING FOR FLOWERS

Within a field of thousands of flowers of the same color, hummingbirds need to remember which flowers they have previously emptied and where they are located, so as to predict when they can visit them again after they have replenished stores of their sweet treats. For a bird that is always on the move but weighs a mere tenth of an ounce (3.2 grams), has a brain the size of a grain of rice, and the metabolism of a hundred ballet dancers, getting a constant amount of enough high-energy food is a difficult problem whose solution has been the evolution of a spatial awareness that is second to none.

ARTIFICIAL FLOWERS IN A FIELD

Rufous hummingbirds are tame but territorial, so remain faithful to a specific foraging patch. Males are constantly on the lookout both for rivals and for females, and feed every ten to fifteen minutes to support their energetic lifestyles. They can be trained to feed from artificial flowers containing sugar water. These flowers are made from the end of a syringe stuck into a cork on a stick. Surrounding the syringe are colored cardboard petals used to suggest different flowers. Typically, an array of eight flowers is presented to the bird. At the start, all the flowers contain nectar, but the birds may only visit four of those flowers. At the second visit, after a five-minute to one-hour delay, all eight flowers are presented but only four contain nectar (the ones not emptied on the previous visit). Hummingbirds were more likely to visit those flowers they hadn't recently emptied.

Below Hummingbirds have a very high metabolism and need a sophisticated spatial memory in order to find flowers that have had time to replenish their nectar stores enough to provide a decent meal until they find the next flower.

COLOR OR SPACE?

What information were the hummingbirds using to remember the flowers? Hummingbirds may favor red flowers as they populate their migration routes, so is color the cue? Four individually colored artificial flowers were presented in a rectangle. One of the flowers contained a sugar solution, but too much of it to be taken in one visit. Once the subject had drunk its fill, the flower was emptied and swapped with another flower in the rectangle. When the bird returned to the rectangle, it was more likely to return to the location at which it had previously fed rather than toward the flower's color, suggesting the birds used location rather than color as their memory cue. Hummingbirds might not use color as a cue because in a field of flowers patches of the same type and color will grow together, so color will provide no useful information whereas location is very precise.

A TIME AND A PLACE TO FEED

When flowers are drained of their nectar, they start the process of producing more. Nectar attracts creatures to the flower that pick up pollen and disperse it across a wide area. It therefore benefits the flower to replenish the nectar quickly, so it can attract its next customer. Do hummingbirds know how long it takes a flower to refill, and if so, can they use this information to maximize their foraging efficiency? Birds were presented with another array of eight artificial flowers, four flowers refilled after ten minutes and the other four flowers refilled twenty minutes after being emptied. The birds revisited the ten-minute flowers sooner than the twenty-minute flowers, remembering not only their location but also when they refilled. This suggests that hummingbirds remember the various replenishment rates of different flowers.

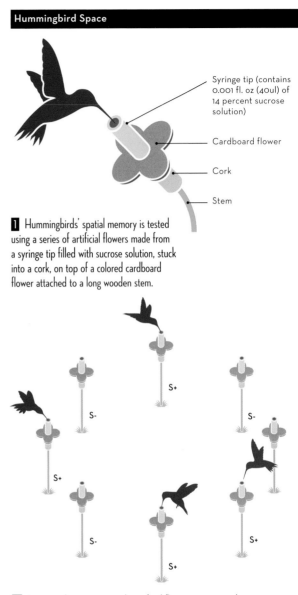

Hummingbird Space

Syringe tip (contains 0.001 fl. oz (40ul) of 14 percent sucrose solution)

Cardboard flower

Cork

Stem

1 Hummingbirds' spatial memory is tested using a series of artificial flowers made from a syringe tip filled with sucrose solution, stuck into a cork, on top of a colored cardboard flower attached to a long wooden stem.

S+
S-
S-
S+
S-
S+
S-
S+

2 In a typical experiment, eight artificial flowers are arranged into an array. Some flowers contain nectar (S+), others are empty (S-). Birds learn the location of S+ flowers.

3 Birds use location rather than color as a memory cue.

An organ for understanding space

The brain area synonymous with navigation is the hippocampus (Latin name for seahorse). The hippocampus has a long evolutionary history, with an equivalent pallial region that processes spatial information to be found in fish, reptiles, birds, and mammals.

STRUCTURE OF THE SEAHORSE

The avian hippocampus consists of two main parts, the hippocampus proper and the parahippocampal area, and is located at the top middle part of the avian brain in both hemispheres, from halfway along the pallium all the way to the back. Although the avian hippocampus looks very different from the mammalian hippocampus, they each receive and project a similar pattern of connections to other brain regions. This looks complicated in the diagram, but it is important to note which areas the hippocampus connects to, how it deals with the information it receives, and how this aids in understanding space. The hippocampus is in a prime position to influence a bird's behavior, because it directly receives information from the senses, connects to regions important for eliciting an emotional response or making decisions, and outputs to regions responsible for producing a behavioral or hormonal response.

NAVIGATIONAL HARDWARE

Most of our evidence for the role of the hippocampus in (non-caching) spatial memory comes from studies in pigeons. When the pigeon hippocampus is lesioned (destroyed), this adversely affects the pigeon's ability to navigate, such as its ability to find hidden food in a maze or form new maps of routes when flying home; but it doesn't affect other forms of memory. In pigeons, the hippocampus has evolved to compute spatial problems, but it is difficult to determine whether it plays other roles in memory without studying other birds.

Neuroscientific studies have revealed a role for the hippocampus in all sorts of navigational problems. Migratory birds have a larger hippocampus than closely related non-migratory species. Although the hippocampus does not function in long-distance navigation, it is essential for registering landmarks when first setting off on a long flight and remembering them when returning home. As such, pigeons may store a landmark-based navigational map of the environment around their home loft in their hippocampus, and use it to guide them back home safely. Brood parasites, such as cuckoos and cowbirds, which lay their eggs in other birds' nests for the hosts to raise as their own, require a specialized form of spatial memory dependent on the hippocampus. Females survey potential nest sites and remember them, before returning when their own eggs are ready to lay, depositing them in the memorized nests with eggs at a similar developmental stage. The female cowbird hippocampus is significantly larger than that of the male, which does not need to remember nest sites. Hummingbirds have relatively the largest hippocampus of any bird, even though the rest of their brain has reduced dramatically in relative size.

NEW NEURONS FOR NEW MEMORIES

The hippocampus is the most plastic region of the adult brain, because it can generate new neurons (neurogenesis), which may aid in forming new memories and updating old information. The relationship between neurogenesis and memory formation is controversial, especially in adult mammals, because it isn't clear whether the new neurons is due to new memories or some other extraneous variable not involved in memory but essential to the learning and memory task. One candidate is exercise, most of the memory tasks given to animals require them to expend a lot of energy. Indeed, migratory birds display more intense hippocampal neurogenesis than non-migratory species, but this might be due more to intense physical exercise than to memory.

Hippocampal Connectivity

A schematic of the avian hippocampus (right hemisphere), seen from the front. Red arrows depict connections with other areas. Purple arrows depict internal connections. (Lower) How information passes through the hippocampus. Light gray (sensory areas), light blue (hippocampus), brown (areas affecting behavior and emotion), green (areas involved in decision-making).

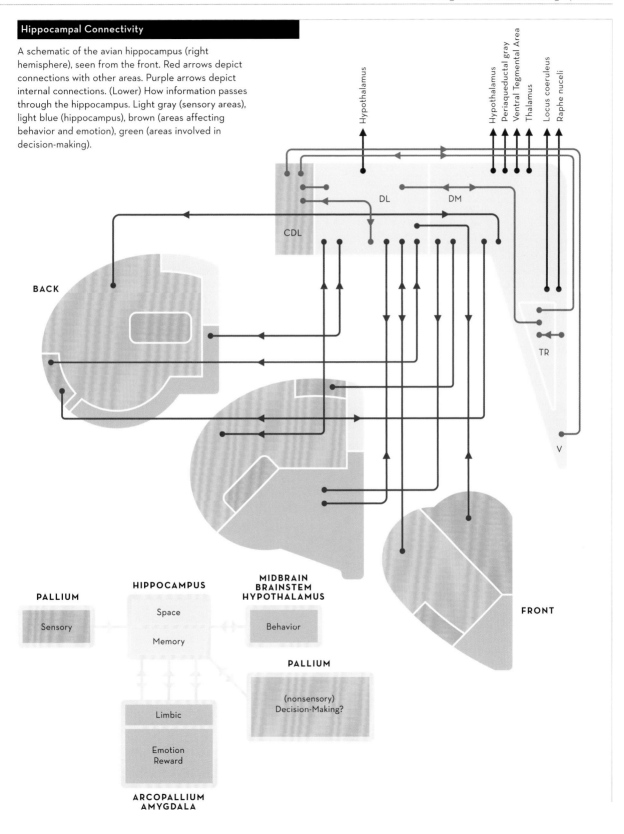

FOOD CACHING

For many birds, the greatest problem is not finding enough food to eat but working out what to do with what they collect. For most species this isn't an issue because they only take the amount needed to satiate their hunger. However, some find enough food for the lean times ahead and hoard it to have something to eat in less favorable future conditions.

SAVING FOOD FOR LATER

For some birds living in a harsh environment, such as black-capped chickadees in the Sierra Nevada, food becomes scarce in the cold winter months. Some mammalian species attack this problem by feasting on enough food to sustain them for the coming months of hibernation. This strategy is not practical for birds, because they cannot lay down the fat stores they would have to rely on during this period. Instead, titmice, such as chickadees, and other birds, cache or hide food in stores.

ONE BASKET OR MANY?

There are two types of food hoarders: larder and scatter. Larder hoarders create one location where they store all their food and protect it. It is easier to defend one central location than many smaller individual stores. Birds that are larder hoarders are extremely rare. Acorn woodpeckers that store their caches in the small holes they create in trees are an example. Birds do not tend to be larder hoarders because they are most active during the day (diurnal), whereas most cache robbers are nocturnal, meaning most birds will not be around to defend those caches. For diurnal birds, it makes more sense to be scatter hoarders, distributing their caches over a wide area and, rather than defending their caches aggressively, protecting them by hiding them out of sight, making them difficult to find.

THE IDIOT SAVANT

The problem with hiding caches from others' view is that they are also hidden from the storer's view, and so they have to evolve a competent spatial memory in order to find them again later. This is not a trivial accomplishment. Some birds, such as Clark's nutcrackers, can hide up to 33,000 pine seeds in the autumn, and have to remember up to 3,000 different locations up to nine months later if there is an especially bad winter at some of the highest elevations. This feat is made especially difficult as the habitat may change between caching and recovery. For

example, snow may have fallen and the landmarks used to form the memory of each cache location may now be concealed. Therefore, these birds cache next to tall landmarks such as trees rather than smaller rocks and shrubs, because they can still be seen after heavy snow.

CUES TO MEMORY

How do birds find their caches? In the two most studied groups, namely parids (titmice) and corvids, it has been found that they do not use odors emitted by the caches (because most birds have a very poor sense of smell). If a cache is smelly enough to guide a bird with poor olfaction, it is likely to be a magnet for mammals with an exceptional sense of smell. They also do not seem to use cues made by disturbing the ground (substrate) in which the caches were buried, as most birds cover their caches well and disguise the fact that they have ever been there.

Birds do not randomly search for caches in the general area in which they hid them, for the reason that this is a terrible technique for gaining the highest rate of cache return and probably expends more energy in searching for the caches than would be gained by eating them. However, they do seem to have a well-developed spatial memory for precisely relocating each cache site as and when required.

Right Shrikes have a novel method for caching food. They eat perishable insects that remain fresh when injured not killed, so by impaling them onto tree thorns, this stops them from wriggling away, and keeps them edible while they collect more food.

Spatial memory, caching, and the hippocampus

Life is tough for a Clark's nutcracker. In order to survive in the snowy tundra of the Rocky Mountains during winter, using their spatial memory they have to rely on accurately finding covered food stores made months before. Nutcracker bills have become adapted to feeding on pine seeds and digging in hard ground. They don't eat anything else, so are dependent on their caches when times are tough.

PLAYING THE NUMBERS GAME

Life is less difficult for pinyon jays, another North American corvid. They also live at a high elevation, but are less dependent on pine seeds, caching around 22,000 seeds a year but also eating and caching other foods. Life is comparably easy for Western scrub jays, the Californian surfer dudes of the corvid world. They live around sea level, stalking the scrublands and parks of the urban western United States, and although they cache, they do so at a much lower intensity than nutcrackers and pinyon jays, at volumes of only around 6,000 pine seeds per year. In laboratory studies of caching and recovery, as well as non-caching tests of spatial memory, Clark's nutcrackers and pinyon jays consistently outperform scrub jays, displaying a reduced number of errors before finding their hidden caches, but also accurately remembering cache sites for longer durations between hiding and recovery. Scrub jays are less reliant on caching, and this is reflected in their poorer spatial memory.

NOT SUCH A SIMPLE PICTURE

This relatively simple picture relating caching intensity, climatic variability, and spatial memory has been presented as one of the clearest examples of an adaptive specialization of cognition, namely a cognitive ability (for example, spatial memory) that has evolved to solve a specific ecological problem (such as

locating caches). However, the picture is much more complicated than it first appears. To begin with, we don't actually know how many seeds each species caches per year, or for how long; the numbers are all estimates, often derived from lab experiments. Second, the focus is on the number of pine seeds cached. Clark's nutcrackers, and to a lesser extent pinyon jays, specialize in eating and caching pine seeds, whereas scrub jays have a wider diet, including invertebrates and fruit berries. Western scrub jays live at a lower elevation than nutcrackers and pinyon jays, yet that doesn't mean they live in a less harsh environment. Having myself lived in the California Central Valley, I can attest that life for a small bird cannot be easy in temperatures of 115°F (46°C) with little water. Scrub jays' preferred foods have a short shelf life, where leaving them for long periods in the hot sun causes them to go rotten very quickly. Such prized foods are also the target of thieves, and so cannot be left for long without protection. So some birds, such as nutcrackers, may have adapted their spatial memory to the problem of remembering a large number of caches, but other birds, such as scrub jays, may have adapted a different form of memory that deals with the different problems of remembering when the food was hidden, what type of food it was and where it was put, and even who or what might have been watching at the time.

Left Black-capped chickadees have a difficult existence, living at high elevations with little food. They have to hide food in the fall for feeding in the long, cold winter months when very little edible grows.

Below American corvids that live at different elevations, and in different environments, have a different reliance on cached food. Species living at a higher elevation cache a greater amount of seeds, than corvids living at a lower elevation.

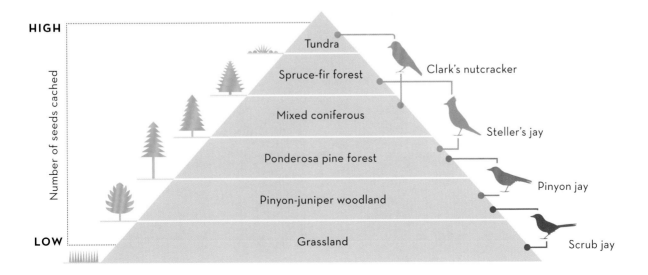

HIGH		
	Tundra	Clark's nutcracker
	Spruce-fir forest	
	Mixed coniferous	Steller's jay
	Ponderosa pine forest	
	Pinyon-juniper woodland	Pinyon jay
LOW	Grassland	Scrub jay

Number of seeds cached

Corvids are not the only birds that cache food. Some woodpeckers, New Zealand robins and shrikes regularly cache, but parids and titmice (chickadees in North America; tits in Europe) are some of the most intense cachers. They do not leave their stores as long as corvids do before recovering them, probably because they are smaller and cannot last as long in cold conditions without regular food. As with corvids, when caching parids such as coal tits, willow tits, and marsh tits are compared to non-caching parids, such as blue tits and great tits, on non-caching spatial memory tasks, the caching species outperform the non-caching species, either remembering where they had seen food or using space rather than color as a cue to finding it.

CACHE LANDMARKS

How do Clark's nutcrackers and black-capped chickadees find their caches sometimes months later? The most popular idea is that they use relationships between cache sites and natural landmarks such as trees and rocks. These landmarks are large enough to be seen when covered in snow, and more landmarks help make the location more precise. Birds use both local and global landmarks. Like pigeons trying to find their home loft, they use larger global landmarks (such as mountains or the tree line) to focus in on the correct region, and then the configuration of smaller landmarks, such as trees, to locate the cache.

As a demonstration of this, nutcrackers were tested in an arena covered with sand, with various rocks located inside. Birds cached food in the arena next to the rock landmarks. In an interval between caching and recovery, one half of the arena and the landmarks on that side were moved 8 in (20 cm) to the right. In recovering their caches, the birds focused those searches on the right-hand side of the arena in relation to the shifted landmarks, not at the actual locations of the caches. On the left-hand side of the arena, they were highly accurate in finding their caches using landmarks which hadn't moved.

MY HIPPOCAMPUS IS BIGGER THAN YOUR HIPPOCAMPUS

Does the hippocampus play a role in caching? Birds that cache rely on their spatial memory more than non-caching birds, and spatial memory relies on the proper functioning of the hippocampus. Despite a long history of controversy on this issue, there is a clear relationship between whether a species caches and the size of its hippocampus. Caching parids have a significantly larger hippocampus than non-caching parids. Food-storing marsh tits have a hippocampus that is 31 percent bigger than non-storing great tits (after correcting for body size). Food-storing black-capped chickadees have a larger hippocampus than closely related, non-storing Mexican chickadees and the bridled titmouse. Among corvids, there is only one known non-cacher, jackdaws, which have a smaller hippocampus than other corvids. Interestingly, European corvids seem to have a larger hippocampus than North American corvids, although

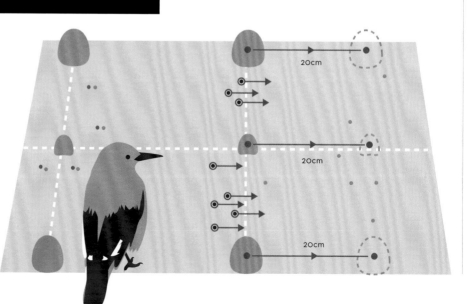

Nutcracker Landmarks

Clark's nutcracker caching seeds in a sand-filled arena including various stone landmarks. After caching, landmarks on the right are shifted 8in. (20cm) and the nutcrackers search for caches based on the new location of the rocks. Blue dots represent the original cache sites on the left, red dots represent the original caches sites on the right, and orange dots represent the search locations.

20cm

20cm

20cm

A Bigger Hippocampus for Memory

Marsh tits store their food, and need to remember the location of their caches in order to use them to help survive a harsh winter. Great tits do not store food, and so don't have this pressure on developing a sophisticated spatial memory. This development in marsh tits is reflected by their 31% relatively larger hippocampus than great tits.

Hippocampus

31%

it isn't known why (though it may be something as simple as differences in the way brains are processed for study by different labs).

USE IT OR LOSE IT

The hippocampus increases or decreases in size (or produces new neurons) depending on whether it is being used. It changes in size both seasonally—with a maximum increase in the volume of the chickadee hippocampus (as well as an increase in new neurons) occurring during October at the peak of food-storing—and depending on need. Populations of chickadees living at higher elevations more reliant on remembering the location of cached food have a larger hippocampus than chickadees living in a more forgiving climate. Development of the hippocampus in young cachers seems to be dependent on a small number of caching and retrieval experiences at a young age, stimulating growth of the hippocampus and an increase in neuron number, but the hippocampus also gets smaller if birds are prevented from caching for only a month. This suggests that the size of the hippocampus, and spatial memory ability, change throughout the year depending on need, so the hippocampus also shrinks in size in the months when it's not needed for caching—a case of use it or lose it!

EPISODIC-LIKE MEMORY

When a caching bird attempts to find its hidden food after a long delay, there is some question as to whether it actually remembers the previous caching event, having tagged the location of the cache site at a specific point in time and space and remembered where it is. The alternative is that the storer has a general feeling of the cache's location, and the contents of the cache, but doesn't remember the caching act itself (i.e. caching is something that has occurred in the past).

MEMORIES OF ROME

This distinction may be subtle, but the first explanation is essential for what is termed episodic memory—remembering specific events in the past that have a location (where), a content (what) and a time the event occurred (when) relative to the current time. This contrasts with possessing knowledge about something, but without that knowledge having personal significance. I know that Rome is the capital of Italy, but I don't remember when or where I learned that fact—I just know it. This is called semantic memory. However, the fact that I remember a wonderful meal on a terrace around the corner from our hotel when my wife and I were on honeymoon in Rome in July 2001 is something that only I experienced and is called an episodic memory. That memory is personal to me, so even though my wife experienced the same meal and the same holiday, her memories of the event will be entirely different from mine, because she will have experienced different emotions from my ones, she will have eaten different food to me, she sat at a different place at the table from me, and so on. Although we shared the same overall experience, our episodic memories will be very different.

AUTONOESIS

At its simplest level, an episodic memory has content about what happened, as well as where and when it did. These have been called what-where-when memories and are slightly different from episodic memories that are the individually experienced memories of our subjective past. Indeed, what-where-when memories do not contain the personal experience of the memory or autonoetic consciousness, an aspect whose inevitable absence has previously made this form of memory in animals so very difficult to study. After all, how do you ask a blue tit how it experienced a past event as opposed to simply what it remembered about that event?

THE EVOLUTION OF EPISODIC MEMORY

The question of whether non-human animals have episodic memories is a very controversial area of science, largely, I think, because episodic memories have become entwined with autonoetic consciousness, suggesting them to be uniquely human. Some important and cleverly designed experiments in a wide variety of animals from cuttlefish to rats and monkeys have found that these creatures can form memories of the what-where-when of an event, such as what type of food can be located (where) in a maze at a specific time (when). This type of experiment doesn't need to refer to consciousness at all. However, a ground-breaking series of studies in food-caching Western scrub jays found they could form memories for specific past events, and importantly that these memories can be flexibly deployed and updated to effect decisions that have to be made in the present.

Right Western scrub jays live throughout the western United States which gets very hot during the summer and very wet during the winter. They need exceptional memories to be able to recover berries and insects before they go rotten.

Memories of what, where, and when

Western scrub jays do not cache large numbers of food items, and do not do so for very long periods. However, they do cache a variety of foods of differing values (meaning that one food is preferred over another) that degrade over time.

FRESH FOOD

Like hummingbirds, scrub jays need an ability to keep track of time, in their case to return to their stashes when the food is still fresh and edible. It would therefore help scrub jays to be able to remember the type of food they cached, where they cached it, and when they cached it, to make sure it is recovered when it is still edible. Nicky Clayton and Tony Dickinson used these facts to design a series of experiments to examine whether scrub jays could integrate what-where-when information together into a cohesive representation of a unique past event, akin to a human episodic memory. To avoid problems interpreting whether the birds were conscious of their own past memories, they named the behavior instead, episodic-like memory.

A TEST FOR EPISODIC-LIKE MEMORY

Scrub jays were hand-raised and divided into two groups: the Replenish Group who never learned that caches go bad, because they were always fresh whenever they were recovered; and the Degrade Group, who learned that some food perishes 124 hours after caching. Jays could cache in an old ice-cube tray filled with sand, made to look different from other trays by the addition of Lego Duplo™ bricks around one side. Birds in the Degrade Group were given one trial caching peanuts in one side of a tray (the other side was blocked), and then 120 hours later they could cache worms in the other, now uncovered, side of the tray. After an additional four hours, they could recover their previous caches. As jays prefer worms to peanuts, they chose to recover worms. In a different trial, they cached worms in one side of a tray, then 120 hours later, cached peanuts in the other side. After an additional four hours, they could recover peanuts and worms. This time, because the worms would no longer be fresh—it was 124 hours since the jays had been cached and they had already learned that worms go rotten after 124 hours—they recovered peanuts, because these were still edible.

Left Western scrub jays live in a variety of habitats, from the temperate forests of the Californian Sierra Nevada to back gardens in Santa Barbara.

In this scenario, the jays could have simply recovered the last food they cached (worms or peanuts) rather than remember the specific caching act. Therefore, jays in the Degrade and Replenish Groups were compared. Both groups cached worms and peanuts in a tray and were then given the opportunity to recover food after four hours. In both groups, the jays recovered worms because these were still fresh. If given the opportunity to recover after 124 hours, the Degrade Group switched to recovering peanuts, whereas the Replenish group recovered worms, because these

were always fresh for them. Therefore, birds in the Degrade Group appeared to appreciate the length of time since caching for two different foods and treated them accordingly, recovering when fresh or leaving them alone when gone bad. As both foods were cached at the same time, the birds were not just remembering the last food they cached. The jays could therefore be said to have remembered *what* they cached, *where* they cached, and *when* they cached it—all components of episodic-like memory, and still the best example in non-human animals.

Episodic-like Memory in Jays test

Episodic-like memory experiment in western scrub jays.

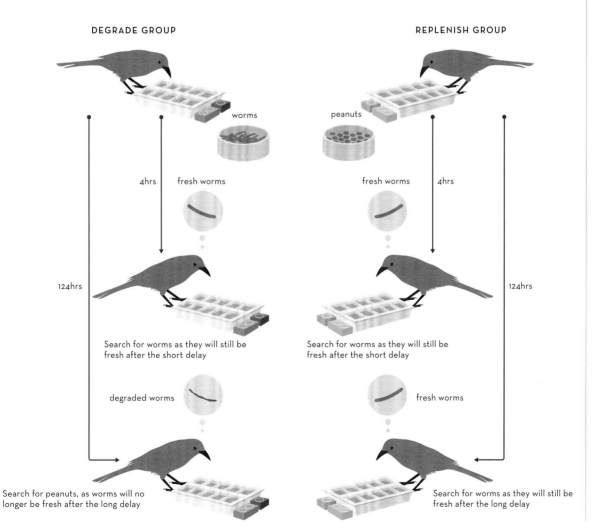

DEGRADE GROUP

REPLENISH GROUP

worms

peanuts

4hrs   fresh worms

fresh worms   4hrs

124hrs

124hrs

Search for worms as they will still be fresh after the short delay

Search for worms as they will still be fresh after the short delay

degraded worms

fresh worms

Search for peanuts, as worms will no longer be fresh after the long delay

Search for worms as they will still be fresh after the long delay

3 GETTING THE MESSAGE ACROSS

How smart do you need to be to communicate?

Communication is a process for transferring information from one being (the sender) to another (the receiver). The sender produces a message containing information in a form that can be decoded by the receiver's brain.

INFORMATION TRANSFER

The information transferred may be simple or complex; it may be transferred between members of the same or different species, such as a prey species' skin color warning a predator to stay away, or recognizing another species' alarm call signaling a predator. The information may be similar independent of species, such as urban birds all producing a 7 kHz alarm call. The information may take different forms, use different substrates (air, water, light/dark, etc.) and span different distances, and change based on background noise and behavioral context. To transmit useful information, the sender reorganizes sensory stimuli into pockets of information encoding an aspect of the world that the receiver cannot perceive without the information being passed onto them.

CONTENT OF THE MESSAGE

Cognition may be required for more complex aspects of communication. Complex messages contain information about specific objects and events, so parts of the message need to refer to something tangible in the world. For alarm calls, a specific sound pattern refers to a certain type of predator, such as an eagle. When the sender sees an eagle, it produces a call using a series of harmonics that says "eagle" to others that can hear the call. Different sound patterns refer to different predators, such as a cat. Unique sound patterns can also refer to different individuals, akin to human names. Learning the relationship between sounds, and what they refer to (referential communication) is an essential step in the development of human language. It is also the step found in those non-human animals that can be

Left Nightingales are one the most melodious birds. They sing during the day and night, but only single males looking for a partner sing at night, with females actively comparing the songs of the nocturnal seducers.

taught aspects of human language. Other components such as recursion and syntax are less common, and the subject of intense debate.

Using information from two different modalities, such as visual and vocal signals referring to the same thing, increases the efficiency of information transfer. Alarm calls tell others there is a predator present, but not where it is. The direction of a bird's gaze informs others where there is something interesting, but not what it is looking at. Perceiving call and gaze direction together increases the chance that receivers understand there is a predator in their vicinity, but also where it is.

INTENTIONALITY

Should we assume that because information is transferred between two or more individuals, the sender intended to communicate? The question remains as to whether the sender intended to change the receiver's behavior as a consequence of the message. Does a male robin squawking at another male invading his territory want him to leave, or want him to believe he will attack if he doesn't leave his territory? Does a sender intend to alert others to a predator's presence because it helps them escape ("Watch out, there's an eagle"), while putting themselves in danger, or is the alarm call just an emotional response to seeing the predator ("Aarrgghh, an eagle!")? If the sender calls and there isn't an eagle, is this because they were mistaken (perhaps misinterpreting rustling in the trees) or because they intended to distract others so they could steal their food? One is accidental, the other intentional, but the behavioral outcome is exactly the same; only what is going on inside the bird's head is different. It is a perennial problem for receivers to interpret communicative intentions in a sender's message.

Avian sensory systems and the brain

The avian brain is finely tuned to processing information about the external world and translating it into plans of action. For successful communication, a bird produces a visual or vocal signal that needs to be interpreted by another's brain as a message requiring a specific behavioral response.

THE SENSORY BRAIN

For all the senses, information concerning the sensory world enters the brain via specific organs such as eyes and ears, which have evolved to convert waves of light or sound energy into patterns of neural activity that can be interpreted as messages and acted upon by the brain.

RED AND YELLOW AND PINK AND GREEN

Birds have more complex eyes than most mammals, reflecting the types of visual information birds need to process, especially during flight. Unlike mammals, bird eyes contain four color photoreceptors, or cones, capable of aiding the perception of a much wider frequency range of light wavelengths than mammals. Most mammals, with the exception of some New World monkeys (tamarins and marmosets), Old World monkeys (macaques, baboons, etc.) and apes (including humans) are dichromatic—that is, they have two cones—and are effectively color-blind, not being able to discriminate green from red. This is not an issue for most mammals, as they are primarily nocturnal, hunting and foraging at night in low light levels. Most primates are trichromatic, with three cones, being able to tell red from green, which is essential for discriminating ripe from not-ripe fruits or between different states of female estrus. Birds, however, are tetrachromatic, having four cones, enabling them to see across the entire visible color spectrum (red, green, blue) as well as colors in the invisible spectrum (ultraviolet) that we cannot perceive unaided. There are important reasons why birds have evolved such a visual system.

EAR, EAR

Birds do not have external ears or pinna, but some, such as owls, have ear feathers that may help to focus sound waves created by their prey into their ear openings (auricles). In owls, one ear opening is higher than the other, so sounds waves entering the ears on either side are mismatched, and can be used to pinpoint moving prey in order to make an effective kill.

PATHWAYS FOR THE SENSES

Waves of light or sound enter the sensory organ (photoreceptor or auricle). The sense organ translates the sound or light energy into an electrical signal that can be detected by sensory nerve cells that transfer this information to the specific sensory pathways of the brain (tectum, thalamus, and then sensory cortex or pallium). The brain translates this information into what is perceived by the bird as a sound stimulus, such as birdsong, or a visual stimulus, such as a face or dance.

Left Owls are one of the most efficient avian predators. Being nocturnal, they hunt at night, relying on their exceptional hearing to locate the tiniest sounds of their mammalian prey. The ability to move their heads in a wide angle, plus their ear tufts, allows them a great accuracy in finding their prey, even when in flight.

Avian Senses

Sight and sound are the main sensory systems of birds. (Left) Diagram of the different cells in the avian retina. (Center) Inside the bird's eye. (Right) Inside the avian ear.

Retina diagram (left):
- Oil drop
- Visual pigment
- Photo-receptors
- Nucleus
- Horizontal cells
- Bipolar cells
- Amacrine cells
- Retinal ganglion cells

Eye diagram (center):
- Retina
- Sclerotic ring
- Iris
- Lens
- Cornea
- Muscle
- Choroid
- Sclera
- Fovea
- Pecten oculi
- Optic nerve

Ear diagram (right):
- Semicircular canal
- Saccule
- Stapes (columella)
- Utricle
- Oval window
- Cochlea
- Skull
- Outer ear canal
- Tympanium

WHY AND HOW DO BIRDS COMMUNICATE?

Birds communicate because they are social and need to transmit information to others. They use a variety of signals in this regard, employing both visual (color, displays, dance, social cues) and auditory (calls and songs) channels to produce the signals and perceive them.

SIGHT AND SOUND

An individual bird is an efficient communicator if it uses appropriate signals in the correct context, as well as the best methods for perceiving those signals. For example, nocturnal birds do not develop visual-based signals, which can only be seen during the day. Birds tend to communicate using either visual or vocal channels, as their olfactory sense is not well developed, aside from those migratory seabirds that use smell to find their way.

Birds communicate information (and withhold information) because it benefits either the sender or the receiver, and usually both. It is often not clear what the sender had in mind when they produced the signal, or whether it even took the state of any potential recipients into consideration. For example, a bird may call to signal a predator. Does this reflect the altruistic intentions of the signaler? Did it produce the call to warn all others, or did it only produce the call for the benefit for its close kin with the knock-on effect being to alert all others to the predator? Or did they produce the call as an emotional reaction with no communicative intention?

SIGNAL PROPERTIES

The visual and vocal signals of birds have evolved to be efficient at transmitting information, and as such are inflexible and cannot be used out of context. A peacock will have inherited his magnificent tail from his father because it helped his father attract a mate and pass on the genes for creating such a beautiful visual signal. Yet the peacock has little control over the size, color, or number of eyespot patterns on his tail and only controls when to display it and to whom. Most signals display four properties. They are stereotyped, meaning that they are usually displayed in the same (predictable) manner; they are repetitive, meaning that they are repeated in order to reinforce

the message; they are simplified, meaning the number of their components is reduced if possible; and they are exaggerated, meaning they are conspicuous and easy to perceive above visual and auditory background noise.

SEX AND VIOLENCE

Birds communicate to transmit information within different contexts, such as to attract a potential mate or warn of a predator. In the context of courtship, birds use a variety of different visual and vocal signals, from colorful plumage and exaggerated ornaments to the construction of bowers or the performance of large repertoires of complex songs. Mated birds use greeting ceremonies, and their contact calls converge to cement their social bond and advertise their relationship to others. Other signals are used to refer to objects in the environment, such as calls signaling the presence or location of food, or alarm calls signaling a predator. Simple signals also dictate normal social interactions. Individuals of a higher social status produce postures reflecting dominance, appearing larger through fluffing up their feathers and standing upright, whereas those of a lower status produce postures reflecting submission, flattening their feathers and crouching down. These seem to be opposite extremes of a similar posture, something Darwin called antithesis. One of the two proposed functions of birdsong is territoriality, where birds advertise their ownership of a specific patch of ground to other males through singing along the borders of that territory. Females use this information to gauge who has the largest territory, and thus who would make a good mate; males need to be successful in order to maintain a larger territory, an advantage that might be reflected in their genes.

Right A male bird of paradise displays his impressively colorful plumage. Only the males are brightly colored and adorned with elaborate feathers, and signal their health and genetic quality to the females, who are usually rather drab in comparison.

VISUAL COMMUNICATION

It can't have escaped your attention that many birds are wonderfully colored. A trip to Papua New Guinea would allow you to experience the greatest range of multicolored plumage in the animal kingdom in the form of birds of paradise. But why such vivid colors when they also announce a bird's presence to predators?

THE FLASHIEST MALE

The simple answer is advertising, not to predators but to potential partners—visual display as a method for attracting the best mate. From a male-dominated human perspective, it may seem somewhat surprising that male birds have the brightest feathers, not to mention the most beautiful voices. But in the avian world it is definitely the males who are the showoffs, and the females who are the drab wallflowers standing in the background. However, because males are trying to attract the females, the latter have all the power, as they get to choose or reject any male that puts himself forward. Moreover, unlike human sexual politics, they don't need to give a reason!

DANCING TEAMS

Other male birds do not develop the natural fashion show of the birds of paradise or some tropical pheasants. Birds such as peacocks produce exaggerated body ornaments with a function that is similar to human tattoos or piercings. In some instances, these extremely long tails or head furniture can make it difficult to fly, even impossible in the case of peacocks with their elaborate and beautiful tails. However, some birds do not develop their bodies in order to show off; rather the way they act is what gets the girls. Manakins, for example, take over seven years to learn to dance in coordinated teams, using lots of different dance styles, but only the dance master, not his apprentices, eventually gets to mate. His trainees only mate once their boss has retired and they start to work with their own dance troupes. Other birds, such as sage or black grouse, dance in a so-called lek, which is an arena in which many males can strut their stuff in front of interested females. Black grouse will bop their wattles, while making short runs, with the movement making a resonating sound. The females choose the male with the most impressive sound and motion combo.

ADVERTISING GOOD HEALTH

One of the problems with being brightly colored or using dance moves as forms of sexual advertisement is that the recipient of your message has to be able to see you. Unfortunately, so can most predators. So why do it? Because their ability to survive despite the danger may be a reflection of good genes and health and is used as a yardstick by which females can assess their quality as mates. A good dancer is also likely to be a good dancer because his father was a good dancer, or because he is in good health. Some traits, such as long or large tails, could be perceived as handicaps, physically preventing those birds from evading predators. Yet females assess these traits positively, figuring that the guy with the large tail that survives long enough to father offspring should have excellent genes to be passed on to his offspring.

Right Royal albatross pairs are strongly bonded, but spend a huge time apart traversing the southern oceans looking for food, though they always return to the same place to breed. When they find their partner again, they re-establish their bond by performing a synchronized ritual dance.

Below A brightly colored male manakin dances for a prospective partner. He will have practiced this dance for years before ever performing it for a female. Despite all his best efforts, if he doesn't achieve the standard the female requires, she'll reject his advances and go looking elsewhere.

The importance of the eyes

Birds, like all vertebrates, have two eyes. Most birds have eyes located at the side of their head, while a few have forward-facing eyes. The different positions tend to reflect whether the species is predator or prey.

EYES FRONT!

Prey need a wide field of view in order to spot an approaching predator, including one coming from behind, whereas a predator has to focus all their attention on the prey in front of them. Prey birds, such as chickens, can detect the presence of eyes (and the number of eyes), because these are a salient clue to the presence or absence of a predator or, more subtly, whether a predator is looking at them (in which case the best response is to flee) or looking away (in which case the best response is to remain still until the predator has left). Predators' eyes are located at the front of the head and, as these birds don't possess complex eye muscles, the eyes don't often move independently of the head; so prey species only really need to be able to discriminate head orientation rather than gaze direction, given that the two are often congruent. A predator moves its head and its eyes follow. A prey species may focus on the presence of the eyes rather than where they are looking. In experiments it was found that sparrows could discriminate between models of humans (as a potential predator) with different head orientations but not different eye orientations within the head.

WINDOWS TO THE SOUL

Eyes provide a more complex signal than whether or not another is looking at you. The eyes may indicate what another is interested in, what they are looking at, and whether they are interested in the same thing as you. What do birds understand about the eyes? In experiments with starlings and jackdaws, a piece of food was placed out in the open and an experimenter faced the food. Across different trials, the experimenter either looked directly at the food or looked away, either with just their eyes or with their head and eyes together or with their head and eyes communicating different things (for example, the head directed away from the food, but the eyes looking straight at it). The birds had to decide when it was safest to take the food. Did they understand that the eyes are the "windows to the soul," that only the eyes are relevant in regard to the act of seeing? The birds could only take the food without being punished if they chose to take it when the experimenter was looking away, such as when their gaze was elsewhere (even if their head was pointing at the food) or both their eyes were closed or their back was turned. They would be punished if they took the food when the experimenter's eyes were open (even just one eye) or looking at the food (even if their head was pointing away).

Starlings and jackdaws distinguished between humans that were facing forwards versus away, or with eyes present versus eyes covered or looking toward food. Jackdaws went one stage further and classified experimenters with only one eye open as seeing the same as experimenters with two eyes open, but seeing differently from one with both eyes closed. Intriguingly, jackdaws only responded to unfamiliar (and so potentially threatening) humans but not to the human who raised and fed them daily, even when they provided similar gaze cues. So the eyes provide clues as to what you should avoid doing and when you don't want to get caught. But eyes can also be essential for learning about the world from others—what they find interesting, where food can be found, and what they plan to do next.

Right There is a wide variety of shapes and colors of different bird eyes. This variety is strongly related to a species' reliance on visual communication, especially color, the food they eat, and the habitats they occupy.

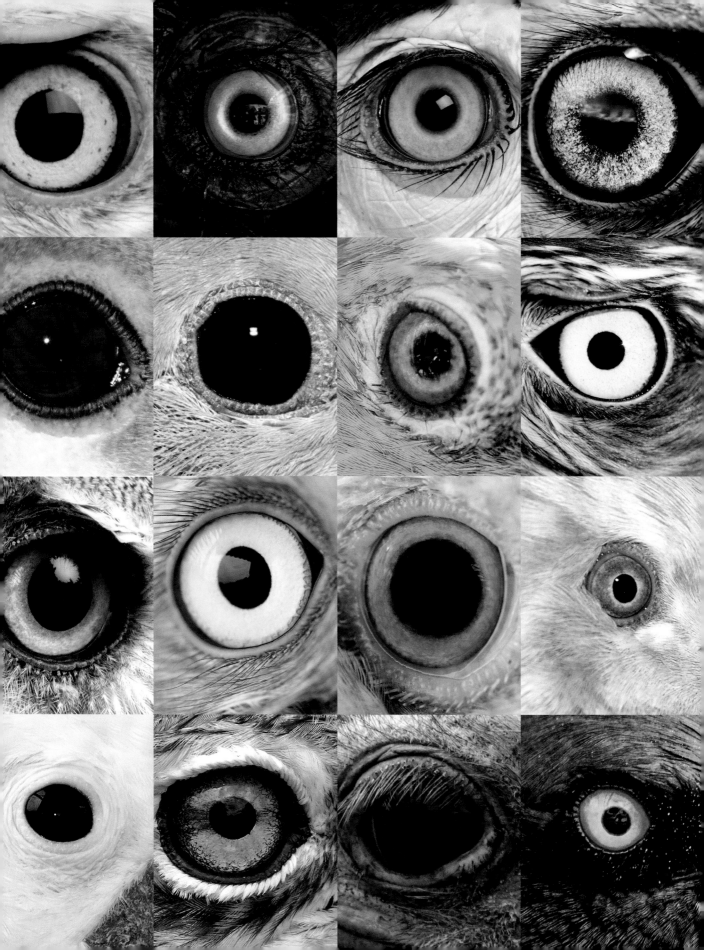

Following another's gaze

Eyes are not just a cue for knowing whether someone is looking at you or for guessing the persons intentions toward you, but also, given that they can look at things, they can be used to learn about objects and events in the world.

WHAT ARE YOU LOOKING AT?

Individuals look at things because things are interesting or because the viewer wants to interact with them. I may look intently at a cream cake because I love cream cakes or because I'm about to grab it and eat it. Equally, I may turn away from something that I find upsetting like crumbs on an empty plate!

THE OBJECT-CHOICE TASK

In a competitive social world, the ability to distinguish objects of interest from another's line of sight—gaze following—is a skill that, perhaps surprisingly, is not possessed by many animals. One of the most common paradigms used to determine whether an individual can understand another's intent based on where they are looking is the object-choice task. Because animals are constantly looking at things in the world (unless their eyes are shut), a viewer trying to gain information from their eye gaze has to distinguish between different objects they might be interested in from among the assortment of random object in their visual path. One way to do this is to focus on the length of time the viewer looks at one object compared to another. A subject is trained to understand that food may be hidden under an opaque box, meaning that although it knows the food may be there it cannot see whether

Below Why should jackdaws be able to recognise the direction in which a human is looking? Like the white sclera of the human eye, jackdaws possess a silver iris that contrasts with their black pupil. This has been proposed as a reason for why humans are so attuned to the eyes of others, and where they are looking, and may also explain the same ability in jackdaws.

the box contains food or is empty. The subject then views either a human or a conspecific looking intently at one of two boxes standing directly between them. The cue the looker provides may be a simple pointing gesture, or she may direct her head and eyes to one of the boxes, or just her eyes alone without moving her head. Only the box indicated by a social cue hides food. When the subject investigates the two boxes, which does it approach first? Surprisingly, most animals that can follow another's gaze, including chimpanzees and ravens, do not choose the correct box at a rate greater than would be expected by chance. Only apes that have been raised in human families, as well as domestic dogs and hand-raised jackdaws, go to the cued box using eye gaze alone.

FOR YOUR EYES ONLY

Only species that have been deliberately socialized or raised with or by humans (or domesticated by humans over thousands of years) pick up intentions from eye gaze. There is some positive evidence for the use of other cues, such as the use of pointing in other species, but this possibly reflects the simpler ability of focusing on the object that is proximal to the pointing hand. Positive cases of using conspecific social cues are surprisingly few, probably because it is difficult to make animals produce a specific cue, and those few cases are restricted to animals that form strong social bonds. Jackdaws pair for life, and it could be advantageous for them to develop the ability to predict their partner's future behavior from simple cues.

Object Choice Task

1 An experimenter looks at or points toward one of two opaque boxes containing food only he or she knows about. A jackdaw evaluates the social cues to choose the correct box.

2 In a second experiment, one jackdaw can see the contents of the boxes and orients toward the one containing food. It is prevented from sitting next to it, being constrained to a small perch. A second jackdaw cannot see the box contents and has to use the other jackdaw's social cues to choose the correct box.

An avian perspective on beauty

Attraction is vitally important to birds. Males spend an inordinate amount of time and effort making themselves as attractive as possible to females. After all, their genetic future depends on it. As such, can we say that birds entertain some concept of beauty or aesthetics?

ART APPRECIATION

Males may develop such an understanding in order to maximize their attractiveness to females, and females may develop such an understanding in order to assess the attractiveness of different males, so as to make the best decision in choosing the optimum (that is, the most attractive) male.

The most likely examples of avian artists are the bowerbirds. Male bowerbirds create complex structures (bowers) usually made from an intertwining of grasses and sticks in vast networks resembling nests that are often adorned with bright, colorful, natural and artificial objects collected together and distributed around and on top of the bower. Males will do anything to ruin the chances of their rivals and will try to destroy other bowers, often distracting the competing males in order to do so.

Male bowerbirds appear to have a particular construction and arrangement of objects in mind when they build their bowers and collect their objects, because they notice when objects they have carefully placed are moved, even slightly, or are stolen by rivals. Females make an assessment of a bower, the collections of objects, and also the presentation skills and vocal ability of the bower owner, before deciding whether to mate with him or move on. Female bowerbirds are perhaps some of the fussiest females in the animal kingdom.

Both males and females could be said to have some aesthetic sense: males in determining what constitutes a bower with the best chance to attract a female, and females in discriminating between different bowers and choosing the best according to whatever criteria evolution has prescribed.

Above Female satin bowerbirds assess the worth of a potential suitor by viewing their bower-building skills. Not only does the bowerbird create an impressive structure, but he also carefully adorns the structure inside and out with various of objects of either the same, or different colors. He is very particular about the placement of these objects in order to enhance their overall aesthetic appearance, and be most likely to attract a female.

JUST AN ILLUSION

Great bowerbirds go one step further in their art appreciation. They create a bower consisting of an avenue of thatched sticks followed by two flat courts covered in both colorless and colored objects. The colorless objects are arranged in order of size, which increases the further away they are from the entrance to the avenue. This appears to cause an optical illusion called forced perspective, which changes the female's perception of the size of the court and the displayed objects by drawing her attention toward the court. Is this art? John Endler has defined art as "the creation of an external visual pattern by one individual in order to influence the behavior of others and an artistic skill is the ability to create art." According to this definition, male great bowerbirds could be seen as artists.

What about the females? Are they art critics rather than artists? Aesthetics is the sense of beauty. It requires a form of judgment or discrimination, as some things will seem more or less beautiful than others to the assessor's aesthetic sense (as they will to the artist). Female animals in many different taxa make judgments about different males, either based on physical traits (such as color, plumage, tail length, or shape), physical ability (for example, dancing, fighting prowess) or the results of the male's physical abilities (bowers and nests). Are these females making aesthetic judgments when they choose based on these traits?

Great bowerbird males create bowers that function by manipulating the female's (choice) behavior—the forced perspective of their avenue construction suggests that they also have an aesthetic sense. Females judge the bowers and discriminate between one and another, so the females might even be said to have a greater aesthetic sense than the males. But we still do not know how the females make these judgments, what are the criteria governing their choice, and whether the choice they are making is geared toward the best chance of subsequent reproductive success. Certainly, in the case of great bowerbirds, the females perceive the visual illusions because they prefer those bowers with forced perspective over bowers without this illusion. Recent findings, however, suggest that the males aren't necessarily being deliberately Machiavellian toward the females. Rather, their choice of objects drives the illusion, with certain objects naturally leading the females to be taken in by the illusion. This would seem to offer some consolation for the females.

Above A male great bowerbird creates a bower by shaping twigs into a long avenue or tunnel, and then lining the outside of the bower with various natural and man-made objects. The objects are larger outside compared to inside the avenue to force an illusion of perspective, so the bower looks larger than it really is.

Communicating with sound

Vocal communication allows you to get your message to as wide an audience as possible. Vocalizations can be used in the dark, are relatively inconspicuous, and can be delivered while on the wing.

SONGS AND CALLS

Birds vocalize either with calls or songs. All birds produce calls, but not all birds sing. Where it becomes a little confusing is who can sing and why. Birds that sing, songbirds, are known as oscines, members of the passerines, the largest family of birds. Young songbirds learn their songs from a tutor, usually the bird's father. Mainly only males sing, because song aids courtship and it is the job of the males to chase the females. However, another group of passerines called the suboscines, including flycatchers, broadbills, and manakins, also sing. They do not appear to learn their rather simple songs but are born with them, but in this group both males and females can sing. To make things more complicated, two other bird groups learn their vocalizations. Hummingbirds sing in a similar way to songbirds, while parrots don't sing, but both males and females vocalize. It has been suggested that they learn their vocalizations in order to learn their partner's contact call (basically their name), and over a brief amount of time the structure of the partners' calls converge into one unique call advertising their partnership and reinforcing their social bond. However, as we will see in the final section of the book, parrots can also use their vocal learning ability to mimic aspects of human language.

A SONG TO SING

Male songbirds learn songs from either their father or a neighboring male tutor. Songs are loud, long, usually with a complex structure, comprising a series of syllables, notes, phrases, and trills of an infinite variety and order. Songs function in territoriality and courtship, so song tends to be seasonal, usually reaching a peak of activity during the breeding season. Males travel around their territory singing, which reduces attempts at invasion from nonresident males.

Left The hyacinth macaw (parrot) is an example of an avian vocal learning species, and has learnt to mimic aspects of human language.

A male that aggressively defends its territory will be seen as attractive to a female. Song also attracts a female, either through its quality—such as the number of different notes—or its length, or the repertoire of songs employed during courtship. The production and perception of song, and specifically the ability to imitate a tutor's song accurately, depends on a complex system of nuclei in the pallium and striatum.

CALLS OF THE WILD

Unlike song, calls are not learned but are used by males and females for many different functions. Calls are normally of short duration with a simple structure. They are used all year round. Examples include food calls, alarm calls, mobbing calls, begging calls, contact calls, and flight calls. While many calls may be an emotional response to something, such as an expression of fear on seeing a predator, they can also be very complex and vary in the amount and type of information they contain.

Although not learned, some calls, especially alarm calls, can be deployed flexibly, with different types of information being recognizable in the calls. For example, African drongos mimic southern pied babbler alarm calls and use them deceptively—that is, when a predator isn't present—in order to steal their food. Black-capped chickadees produce more calls when they perceive more dangerous predators, such as small owls; smaller predators are more deadly to chickadees. Arabian babblers increase their calling rate when the threat from a predator is greatest—that is, when they are moving closer. Hornbills recognize the eagle alarm calls of Diana monkeys, but ignore Diana monkey leopard alarm calls, as leopards are no threat to them. Finally, chickens can discriminate different predator types based on the structure of the alarm call. When presented with a ground predator alarm, chickens frequently scan the ground, whereas when they are presented with an aerial predator alarm call, they scan the sky.

STRUCTURE IS RELATED TO FUNCTION

Information in the call is related to its function. European blackbird mobbing calls are short duration, with a broad frequency range, so others can hear them and become recruited as such calls make it easier to locate the caller. Blackbird alarm calls are longer duration but also a steady frequency around 7kHz, meaning the duration is long enough to be useful in communicating a message but is more difficult for a predator to use in pinpointing the caller's location. Predators cannot hear this frequency, which is why many birds use alarm calls around 7kHz.

Animal Hearing Ranges

Bird calls and song are within human hearing range, whereas many insects, bats and cetaceans communicate with sounds outside of our auditory perception

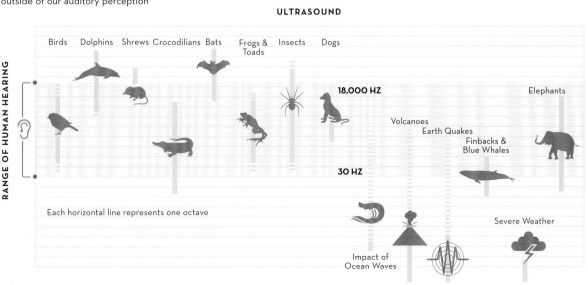

Singing the same old song

Calls are inherited not learned. If a chicken is raised in isolation or deafened before it hears its first call, it will still produce calls as an adult. Suboscines that sing but do not learn their songs will sing normally as adults even if prevented from hearing another conspecific sing.

A SENSITIVE PERIOD

Oscines and hummingbirds will not sing if they are prevented from hearing a song model, usually their father, during a sensitive period in their development. This used to be called the critical period, suggesting a level of inflexibility, though we now know in some species that the timing of song learning is not as strict as previously assumed. Canaries, for example, have no sensitive period as they will continue to learn new songs throughout their lives.

FATHER'S SONG

There are two phases of song development: the sensory acquisition phase and the sensorimotor phase. In the sensory acquisition phase, the young bird hears a model, usually produced by their father.

Following this phase, there is a silent period during which the song learner memorizes the pattern and structure of the tutor's song, but does not vocalize. Young birds possess an innate song template containing the basic elements of their species' song. A surge in testosterone initiates the sensorimotor phase, during which the song learner practices its new song (sub-song), comparing it to its innate template through auditory feedback. It remembers the learned song, sings notes from its innate template and matches those to the remembered song. It continues to refine its plastic song, retaining elements that match and discarding elements that don't, until its song is a direct copy of the learned song (crystallized song). Some species, age-limited learners, will only learn this one song, which has to

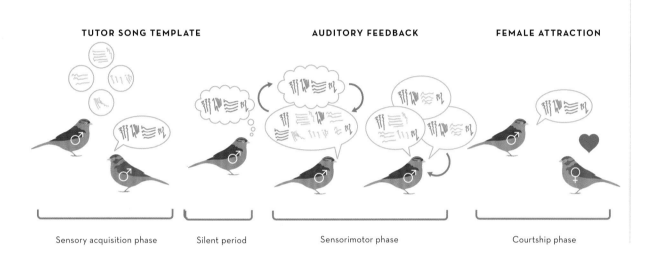

Song Learning

An illustration representing the different phases that occur when a young male songbird successfully learns to reproduce a song from a tutor and use it to court a female.

TUTOR SONG TEMPLATE   **AUDITORY FEEDBACK**   **FEMALE ATTRACTION**

Sensory acquisition phase   Silent period   Sensorimotor phase   Courtship phase

occur during the sensitive period. The length of time the learners have to learn differs between species. White-crowned sparrows learn when between ten and fifty days old, whereas chaffinches are receptive ten to twelve months into the first breeding season. Young birds that do not hear song during the sensitive period do not develop a proper song, but they will sing. Their song is less complex, contains fewer notes, displays fewer changes in frequency, and is less attractive to females. Other species, such as lyrebirds and mockingbirds, are open-ended learners with no sensitive period, and adopt other sounds and calls into their vocal repertoire, such as those of other species and the sounds of manmade objects.

VOCAL CULTURE

Birdsong varies within a species depending on geographical location. This variation is more likely to occur if there is some natural barrier between

populations to prevent intermixing. Because birds copy tutors they are related to or familiar with, there are few opportunities for diversifying song structure; so by restricting song to those in the same vicinity, regional dialects form in a similar way to local accents in human populations. Differences may be structural or in the way that a song is produced, such as its speed. The classic case for song dialects is white-crowned sparrows in the San Francisco Bay Area studied by Peter Marler. Dialects function in defending territories, as males act more aggressively toward other males producing foreign dialects than they do to their neighbors. By copying the song of older but still sexually productive males, young birds may be carrying on a tradition in the form of a song dialect.

Songs may also be classified as cultural traits that are passed along from father to son (vertical transmission) but also within a generation (horizontal transmission) if passed between neighbors. Like human cultural traditions, the core song structure is retained, but some novel additions will naturally be added across generations.

Left Canaries are one of the world's most voracious avian singers. In the 17th century, they were highly prized by European monarchs, but rapidly became a commonplace pet. Canaries were one of the first species to be studied to help in understanding the neural basis of birdsong.

The avian song control system

A neural circuit controls the perception of songs and calls, memorization and imitation of these vocalizations, and their production.

CHATTER BRAINS

Songbirds, parrots, and hummingbirds all possess a form of this circuitry spanning nuclei in the pallium and striatum, linking inputs from the auditory system with motor outputs via the syrinx (the avian equivalent of the human larynx). Although each group has a differently structured brain, they share functionally equivalent regions for vocal learning.

Non-learning species, such as chickens, do not possess these regions and have a much simpler circuitry for producing their calls. In the artwork opposite, the brains of a mynah bird (songbird), an Amazonian parrot, and a hummingbird display similarities in their vocal systems. Nuclei with the same color form part of the same pathways across birds.

SONG CIRCUITRY

In songbirds, three neural networks connect the perception, memorization, copying, and production of song. As males sing during the breeding season, we expect there to be sex and seasonal differences in the relative size of the song circuit. The song circuit is much larger in males; the higher vocal center (HVC) is three times larger in male canaries and eight times larger in male zebra finches. It is also larger in the spring than in the late summer. New neurons are born in the spring and are constantly replaced during the breeding season. Canaries produce a new repertoire of songs each year and develop new neurons in the HVC each year. It is clear, therefore, that the song circuit is as plastic as the hippocampus.

In the sensory acquisition phase, song information enters the brainstem, thalamus, and Field L (primary auditory area) via the secondary auditory regions: the caudomedial nidopallium (NCM) and mesopallium (CM). This is where song is perceived as song (compared to other sounds). The tutor's song is recognized, and discriminated from background noise and other songs so that the correct song is copied. Secondary auditory regions are not considered part of the song control system, because they are not dedicated merely to song. Songbirds, parrots, and hummingbirds all share these regions, and these structures have been conserved since the time of their last common ancestor. Memory of the tutor's song is retained in the secondary auditory regions. Neurons in these regions show increased responses to song stimuli that a bird has experienced before. Blocking receptors in the NCM normally active during song learning has the effect of preventing the tutor song being stored there, leading to the development of an abnormal song.

In the sensorimotor phase, as the young bird starts practicing—like a human infant babbling—the song motor pathway is activated. This pathway includes the HVC, which receives information about the learned song from memory, and starts to produce sub-songs from the bird's innate song template. The HVC projects to the RA nucleus, which in turn activates the motor system controlling the movement of the syrinx and breathing during song production. The bird produces the beginnings of a song, hears this song, and using auditory feedback corrects any errors in relation to the memorized song.

A final anterior forebrain pathway is essential for translating the elements of the song that the bird perceives into a copy of the memorized song. By refining innate elements of song (plastic song) and comparing them to memorized song via a looped pathway of the HVC, Area X (basal ganglia), thalamus, LMAN (anterior nidopallium), and RA, the bird produces an imitation of the original song. The LMAN to RA pathway is particularly important for this, as LMAN lesions stabilize plastic song rather than turn it into crystallized song, so the young lesioned birds will be frozen in their immature state, producing nothing more than disjointed song elements. Once the song has crystallized, the bird produces song elements in a stereotypical sequence closely resembling the tutor's.

Neutral Basis of Song Learning

The diagram of the oscine song system shows blue arrows representing connections of the secondary auditory regions (hearing song). The red arrows represent connections of the anterior forebrain pathway (imitating song). Green arrows represent connections of the song motor pathway (producing song).

(Below) The brains of three vocal learners (songbird, parrot, and hummingbird) and a non-vocal learner (chicken). Red nuclei are part of the anterior forebrain pathway, green nuclei are part of the song motor pathway, whereas blue are secondary auditory regions. Chickens only have parts of the motor pathway that play a role in vocalizing.

VOCAL IMITATION

About 20 percent of songbirds are open-ended learners with no restriction on when they learn new songs. Some species learn entirely new repertoires of new songs each year. Classic cases of vocal imitation or mimicry are mockingbirds, starlings, mynah birds, and lyrebirds, which imitate huge numbers of songs but also other species' calls and even human machinery.

COPYCATS

Mimics are often so accurate they fool humans. Yellow-billed magpies used to frustrate a friend by causing her constantly to run for a ringing phone that never actually rang. There seems to be no limit to what some imitators can add to their repertoires. Mockingbirds, for example, have 150 songs in their songbook, a number that increases with age and includes frog, insect, and bird vocalizations as well as different alarms. Vocal mimics that migrate pick up calls from an exotic range of species. Marsh warblers mimic around 76 species, with 60 percent of them African, living around their migratory breeding grounds. Parrots are the best-known mimics, and have been exploited as pets because of their ability to imitate human speech. Think how different our relationship with dogs would be if they could talk back to us?

WHY MIMIC?

There must be an important evolutionary reason why some birds expend the energy required to mimic such a wide range of sounds with such accuracy. One suggestion is that males who produce a wide repertoire of sounds, including nonvocal sounds, and display great prowess in accurately copying the sounds are more attractive than males who only copy a limited number. This would be no different from the male satin bowerbird that collects a wide range of blue objects, some difficult to get. Male satin bowerbirds with the largest sound repertoire and the greatest level of copying accuracy were also the males with the highest level of reproductive success (the most attractive). Another suggestion why birds imitate is to mimic predators so as to distract others from goals such as food. Bowerbirds will mimic a predator to scare off a rival, so that they can then destroy its bower and court

females with decreased competition. Parrots may have evolved mimicry to bond with their social partner, as the male and female eventually manipulate the structure of their individual contact calls into a single paired contact call. Parrot owners can exploit this fact when they teach their pet parrots human words; the parrot may learn the words faster if they have developed a stronger bond with its owner.

DANCE TO THE BEAT

Vocal mimics possess an additional skill that could be relevant to understanding the evolution of human music and dance. When a piece of music with a beat is played to vocal mimics, they quickly become entrained to the rhythm, moving in time to the music. A sulphur-crested cockatoo called Snowball became an internet sensation in 2008 when a video of him dancing to a track by the Backstreet Boys went viral. He accurately followed the beat of the track, bobbing his head up and down, moving his feet alternately, raising his head crest and adjusting his movements in concert with any changes in tempo. In a wide-ranging study across birds, vocal mimics were more likely to entrain to a beat than non-mimics, yet it's not clear why vocal imitation is essential for this skill.

Above Mockingbirds are renowned for their ability to mimic all sorts of natural and unnatural sounds, including the calls and songs of other birds, and manmade sounds, such as car alarms.

Right A lyrebird is one of nature's most impressive vocal mimics and have been found capable of copying camera shutters, chainsaws, car alarms and a variety of different species' bird calls, such as kookaburras.

TEACHING A PARROT TO TALK

The pallial vocal learning pathways of parrots, which are functionally similar to songbirds, provide a supportive structure for them to learn human language. Indeed, the ability to mimic speech has been termed "parroting."

PARROTING

Many with pet parrots marvel at the number of human words their feathered friends can speak, sometimes even achieving conversations appropriate to their context. Do some parrots do more than just "parrot?" The most famous example of a parrot doing more than mimicking human speech was an African grey parrot called Alex, who was trained and tested for 30 years by Irene Pepperberg. Unfortunately, Alex died in 2007, but studies on his understanding of language were some of the first to highlight the complexity of the avian brain and cognition.

A PARROT CALLED ALEX

Alex was chosen at random from a Chicago pet shop. Pepperberg was interested in whether parrots could be taught to speak, but she was particularly keen to use language as a tool for investigating a bird's understanding of concepts, such as number, how objects can be categorized, and whether objects share features. It would certainly be easier to ask your subjects questions rather than having to design complex experiments. Alex was taught human words and their meaning using the Model/Rival technique, in which two trainers interacted while holding an object, and one of them (trainer) questioned the other (model/rival) about the object (for example, "What's here?", "What color?"). When the model produced the correct response, he or she received praise and the object to play with as a reward. Incorrect responses, such as producing a deliberately garbled or incorrect response, received a scolding and the object was removed. Alex was also questioned and rewarded for approximating the correct response. Over time, what constituted a correct response became narrower so that Alex eventually learned the correct word. During testing, the trainer and model/rival switched roles across sessions, preventing Alex from focusing only on the trainer, rather than the model/rival.

WHAT DID ALEX LEARN?

Eventually, using this method Alex learned to respond correctly to the names of more than 50 objects, seven colors, five shapes (based on the number of corners), quantities up to eight, three different categories related to objects (color, shape, and material), simple commands such as "No," "Come here," "Wanna go X" or "Want Y,", as well as more abstract concepts such as whether two objects were the same or different (based on abstract properties, not just how they looked), and whether they differed in size and/or number. Alex also learned to combine labels, which helped him to identify, request, comment on, or refuse objects.

DID ALEX UNDERSTAND WHAT HE WAS SAYING?

When Alex was presented with an array of objects, differing in color, shape, or material (or having shared features), and asked questions about those objects, he was very accurate in providing the correct answer. Alex correctly answered questions such as "What object blue?", but also questions requiring understanding across categories, such as "What matter four-cornered blue object?" or "How many blue objects?" Some objects shared the same property, such as color,

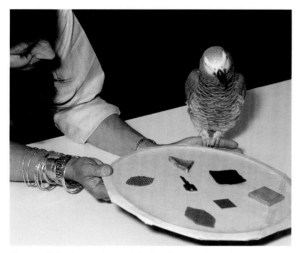

Above and left Alex was also capable of performing other tasks requiring an understanding of different concepts, such as discriminating between different objects based on various properties. Alex (left) was asked to count the number of objects of the same color and (above) "what matter four-cornered blue?" meaning "what material is the blue square made from?" Alex was always very accurate in his responses.

but not others, such as shape. Differentiating the number of blue objects in an array (for example, the blue circle and blue square = 2) or the four-cornered objects (for example, blue square and red square), requires a conjunction of object properties (that is the multiple properties objects possess that are used to tell them apart from other objects that share some of the same properties). Alex therefore seemed to understand his utterances, using them appropriately, and using some words spontaneously outside of the context in which they were learned. But was it human language? Certainly, Alex produced and understood labels, concepts, actions, and relations between objects, so he can certainly be said to have communicated symbolically, a core component of human language. But language is more than just symbolic communication, utilizing grammar and syntax to support the meaning and intention behind any utterances. So far, only a couple of language-trained bonobos and dolphins have demonstrated this level of language comprehension.

4 FEATHERED FRIENDS (AND ENEMIES)

Why do birds live in groups?

Most birds live in groups, but there are both costs and benefits in doing so. Living with others can enhance your security against predators, but by increasing the size of the group you also make it more visible. Indeed, increasing group size gives potential predators a wider choice of prey, so being stronger than the next guy also increases your chances of survival.

SAFETY IN NUMBERS

It's bad luck if you live at the edge of the group, where you're more likely to be taken, but good news if you occupy a central position; naturally, these positions tend to be occupied by those with the highest status. Producing offspring within the confines of a group affords them more protection, but it also increases the chances of alerting a predator to their presence. However, group living also increases the chances of spotting a predator and letting others know about it through the use of alarm calls. For a bird faced with a predator hunting by surprise, alerting others to their presence by raising an alarm can mean surviving another night. Some social birds, on the other hand, even those of different species acting collectively, have developed more aggressive survival strategies and will gang up to mob any uninvited diner with enough ferocity to put them off.

OPPORTUNITIES

There are other advantages to living in a group, such as increasing opportunities for learning from others, for locating food or sharing the methods for extracting food found inside hard cases or in awkward places. Birds seeking food that is only available seasonally or in specific locations will gather in great numbers and inevitably attract predators. Moreover, food concentrated in one place for a short period also leads to other problems, such as having many mouths to feed at one time, which can often result in fighting.

It is easier for social birds to find a mate, which is especially important if they mate with multiple partners across a breeding season. Even if birds only mate with one partner, as in the case of the royal albatross, they still live in large flocks. Living with others in a harsh environment, such as the Antarctic, can be life-saving given the opportunities for huddling together to increase warmth; Emperor penguins are well-known for this behavior.

EVOLVING A SOCIAL INTELLIGENCE

There are other disadvantages to social living aside from sharing resources or advertising dinner, among them the tendency to increase the potential spread of disease or parasites, both of which occur more frequently in close proximity to carriers. These hazards aside, social life is not easy, but given that most birds are gregarious rather than solitary, it seems that birds in general are capable of overcoming any difficulties. Most species sort out their problems by knowing their place in the social hierarchy, and most get along without the need for any special cognitive skills adapted to the unpredictable social world. However, a small group of birds may have developed a form of social intelligence at one time assumed to be found only in primates.

Scientists believe that social rather than physical problem-solving was the primary driver of human cognitive evolution, given that the social world can fight back. Dealing with social agents, with their own personalities, relationships, likes and dislikes, intentions and plans, adds a level of complexity not found in dealing with physical objects. Until the 1960s, it was assumed that only humans required complex social skills. Then Jane Goodall and Dian Fossey reported startlingly similar social sophistication in our closest ape relatives. Now we know that these skills are found in a wider range of vertebrates than was imagined 60 years ago, including many birds. In the rest of this chapter, I will describe what is known about avian social intelligence, considering the skills they may share with mammals as well as traits that make them unique.

Left Emperor penguins live in huge colonies, occasionally numbering more than 1 million individuals. They congregate together for warmth in areas rich in fish, but each bonded male and female pair is only capable of raising and protecting one chick because this environment is so harsh.

The birds and the bees

For most animals, getting and staying together has only one purpose: the production of healthy offspring. Any associated benefits from safety in numbers, increased foraging opportunities, or learning something new remain circumstantial.

THE MATING GAME

The chance of finding a suitable partner increases within a larger social group; in addition there is a better chance of producing offspring that will survive into adulthood. This idea forms the basis for promiscuity, a system in which individuals of either sex mate as many times and with as many individuals as they can. In a polygamous system, males mate with as many females as possible during the breeding season, theoretically producing as many offspring as possible with many mating partners. The males mate and go, never taking part in raising their chicks; however, most birds do not do this. A tiny percentage of birds are polyandrous, indicating a female mating with multiple males at the same time (so the reverse of polygamy). The males not the females raise their chicks, because once the female has laid her eggs, she leaves. The majority of birds, however, are monogamous, using a system in which a single male mates with a single female during the breeding season and contributes to raising his chicks, helping the female with feeding and caring for them. Some birds mate for life, taking a partner at a young age and sticking with them for the rest of their lives.

SEXUAL CONFLICT

Some species, such as dunnocks, continually change their mating system; most frequently they are monogamous, but they are also polygamous and polyandrous when needed. Males and females benefit from different things, so pursuing them may lead to conflict. Overall, females are better served if their partner helps raise their offspring, because this reduces stress on their bodies and increases the chicks' survival prospects. If a male helps but has produced offspring with a number of other females, then his time will be divided, cutting the time he can dedicate to any one female and her brood. Therefore, monogamy benefits the female but not necessarily the male. Polygamy benefits males because they can mate with a large number of females and then leave, thereby ending their parental role. Many of his offspring will die without his help, but creating more chicks overall may lead to more of them surviving, and all for little personal investment on his part. Compare that to a monogamous male who devotes all his time and energy to a small brood that could all be taken by a predator anyway—putting all his eggs in one basket!

DECEPTIVE DUNNOCKS

With dunnocks, the battle between males and females becomes most pronounced when individuals use deception to maximize their benefits. A dominant male and female mate, but after the male has left, a lower-ranked male may try his luck mating with the female, and she may let him. Sex by dunnocks is brief, and sperm is deposited quickly between cloacal openings (male dunnocks do not have a penis). After sneakily mating with this lower-status male, who has now left the scene, the female may advertise to her mate that she has mated with another by wafting her tail feathers in his face, and he will forcibly peck out the usurper's sperm from the female's cloacal opening before mating with her again. Both males have mated with the female, but only the dominant male will be the father of her offspring. Both males will help raise the offspring, increasing their chances of survival; however, one of the males has been conned, and is investing time and energy in young that do not share his genes. He would have been better off playing the field.

What isn't clear is to whether birds need social smarts in order to go through these seemingly complex acts of deception? The female's behavior is the result of millions of years of evolutionary tinkering, and occurs whenever certain circumstances are present rather than through the individual female thinking through her options. It may be possible to determine the extent of the female's flexibility and intelligence using experiments, but these have yet to be devised.

Avian Social Systems

1 At the top of the diagram, birds live in larger groups or colonies with overlapping territories, primarily for protection from predators. Ravens (top left) represent the classic social system in birds in which pairs (P, darker circles) live in close proximity, with non-overlapping territories (lighter circles) containing non-breeding non-helpers (NH).

2 Blue jays (top middle) show the beginnings of colonies with pairs living in the same territories as other non-breeders close by.

3 Jackdaws (top right) are an example of a large nesting colony, with many pairs living in the same territory (that is, the overlap is too significant to enable us to distinguish between different territories).

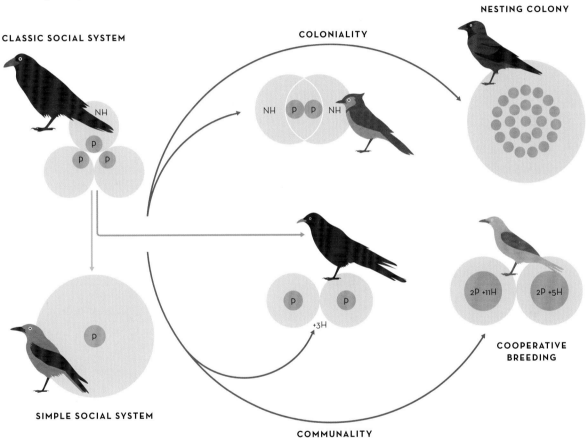

NESTING COLONY

CLASSIC SOCIAL SYSTEM

COLONIALITY

NH P P NH

P
P P

SIMPLE SOCIAL SYSTEM

COMMUNALITY

P P
+3H

2P +11H 2P +5H

COOPERATIVE BREEDING

4 At the bottom of the diagram, birds live in smaller communities with distinct territories. The simplest form is represented by Clark's nutcrackers (bottom left), in which a single pair breed in a large territory, remaining practically solitary outside the breeding season.

5 Alpine choughs (bottom middle) also live primarily in pairs, but environmental conditions may necessitate pairs living in simple communities, with non-breeding helpers.

6 Finally, Mexican jays (bottom right) represent the case of a true community—cooperative breeding—in which multiple pairs live together with their offspring (non-breeding helpers, H) from previous years, who help in raising their current brood. The number of helpers will depend on the quality of the current environment, which may prevent the younger birds from raising their own families.

The avian social brain

Birds face environmental challenges that can be categorized as social (dominance, reproduction) or non-social (foraging, habitat selection, predation), and have evolved neural systems to deal with such challenges and create opportunities.

All vertebrates' brains process information about an animal's immediate environment that enables the animal in question to make appropriate, often life-saving decisions. Which fruit should I eat? Which female shall I try to court? What is the best way to avoid being eaten? For an animal living in a group, such decisions are some of the most important and most difficult to make, because they are directed toward beings capable of making their own decisions.

A SHARED PLAN FOR SOCIALITY

The overall social brain plan is the same across all vertebrates. Information about a social stimulus, such as its appearance, smell, or sound, enters the brain via those sensory organs appropriate for a given species—primarily the visual and auditory systems for birds and primates; olfactory system for rodents and so on). This information is analyzed until it can be categorized as a face, pheromone, or call. But, whose face, pheromone, or call? What is the relationship to you, the social status; has the individual helped you in the past and do you need to return the favor? Therefore, for a creature with social knowledge it is not sufficient simply to recognize a social stimulus, but to evaluate it for its salience, emotional value, or social significance and initiate an appropriate behavior response. For example, an individual directs a threatening facial expression toward you. That individual is dominant to you. This requires a certain behavioral response in order to prevent aggression toward you, such as displaying a submissive gesture. If a subordinate directed the threat, then the appropriate response should be different, such as a return threat or even attack.

The vertebrate brain includes two primary networks as part of a social evaluation system, the Social Behavior Network (SBN) and the Mesolimbic Reward System (MRS). The SBN consists of a circuit including the hypothalamus, medial amygdala, and periaqueductal gray (red) which collectively are involved in evaluating the emotional significance of a social stimulus, especially in the context of sexual behavior and initiating a response via hormones. If a female invites a male to copulate, then the SBN will be involved in translating her invitation display into the correct actions for copulation. The MRS consists of a brain circuit including the hippocampus, striatum, pallidum, and basolateral amygdala (blue), that are involved in the reward system. In the context of social behavior, this system could be important for forming social bonds, such as remembering a long-term pair mate. "Two neuropeptides in birds, vasopressin and mesotocin, and another, oxytocin, in mammals are essential transmitters in this system, and their levels in this network are found to be higher in species with strong social bonds. In our example, a male may become bonded to a female after repeated matings due to the activation of this reward system. The lateral septum and bed nucleus of the stria terminalis (purple) are part of both the SBN and MRS, perhaps forming a connection between them.

THE SOCIAL KNOWLEDGE NETWORK

These two brain systems have remained stable across millions of years, because they are present in all vertebrates, from fish to mammals. Yet there is little to suggest that they play more than a minor role in social knowledge. It is therefore possible that those species that demonstrate repeated social interactions within a stable group require an extra neural system for processing social stimuli, storing those stimuli as categories and retaining social memories for a long time. Such stores could include others' relationships with others (dominance and affiliation), cooperative book-keeping (such as who owes what to whom), memories of past interactions, and a system for predicting another's intentions. This social knowledge network is currently theoretical, but preliminary studies have suggested it may include higher-order areas of the avian brain such as the entopallium, mesopallium and nidopallium.

Avian Social Brain

The brain below displays the connections of the regions of the social brain. Areas in blue are part of the Social Brain Network, important for attributing emotional significance to a social stimulus, whereas areas in gray are part of the Mesolimbic Reward System that attributes value to a stimulus (reward). The areas in purple are part of both circuits and may form a connection between these two systems. The illustration on the right shows the main processing steps involved in social decision-making (perception to evaluation to response).

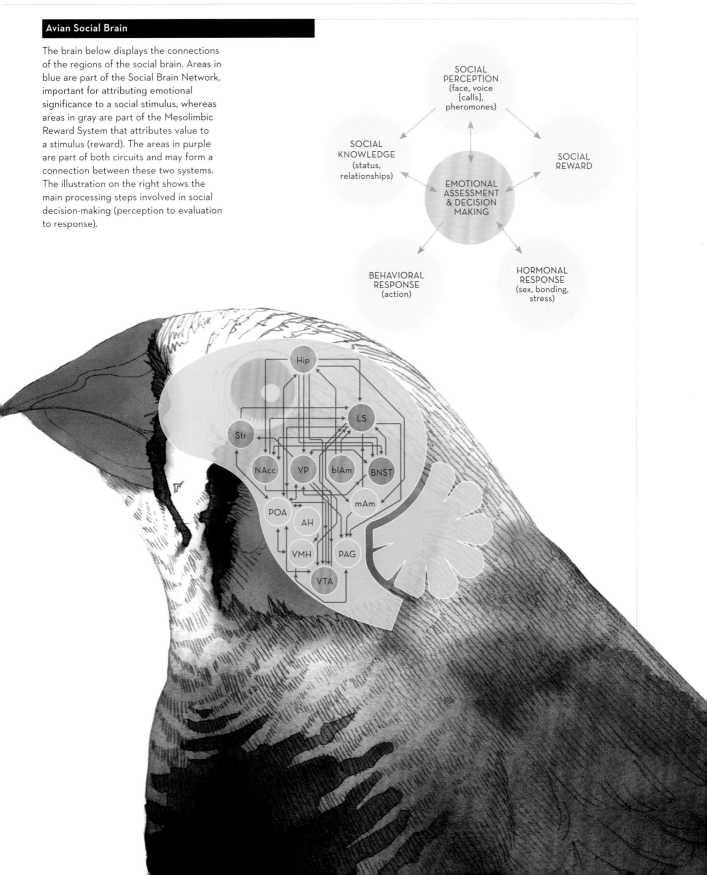

SOCIAL PERCEPTION (face, voice [calls], pheromones)

SOCIAL KNOWLEDGE (status, relationships)

SOCIAL REWARD

EMOTIONAL ASSESSMENT & DECISION MAKING

BEHAVIORAL RESPONSE (action)

HORMONAL RESPONSE (sex, bonding, stress)

Hip

LS

Str

NAcc   VP   blAm   BNST

mAm

POA

AH

VMH   PAG

VTA

SOCIAL CLIMBERS

Living in a large group, how do you decide who gets what and in what order? One option is to fight. Individuals differ in size, speed, and fighting ability, and this contributes to dominance. Unsurprisingly, those with the highest dominance tend to secure the best resources.

KNOW THY PLACE

The Norwegian zoologist Thorleif Schjelderup-Ebbe conceived of dominance after observing a flock of chickens. He found that two birds, when given food, would fight over it, with the most aggressive bird, delivering the greatest number of pecks to secure the food, using violence to get what it wanted. A new pair would go through the same process until each bird could accurately assess its chances of winning or losing a fight. From then on, actual aggression was not needed to maintain the pecking order, because the threat of aggression based on a protagonist's dominance status was now sufficient. Dominance allows individuals in a social group to sort out who gains access to resources without resorting to risky fights.

Individuals are sorted into a dominance hierarchy with the most dominant (that is, most aggressive) at the top and the most subordinate (that is, most submissive) at the bottom. In the hierarchy, the response to another's peck is either to fight back (if the pecker is lower status) or to produce a submissive gesture, such as lowering the head (if the pecker is higher status). A dominant individual can therefore threaten aggression without attacking, and a subordinate one can demonstrate submission without having to be the victim of aggression. By adopting this code, individuals "know their place" and can live in relative harmony.

Pecking rates can be used to construct a stable dominance hierarchy with bird A pecking at the highest rate, and other members being lower in the linear hierarchy, so that A > B > C > D > E. By knowing each bird's place in the hierarchy and each individual's own dominance rank, it is possible to determine who is dominant of whom, even if the two individuals have never interacted before. As B is dominant to C, and C is dominant to D, if B and D meet, then D should produce a submissive gesture to prevent B from attacking. An ability called transitive inference is said to underlie this understanding.

Physical aggression is not the only measure of dominance. Size matters. Larger animals tend to be more aggressive and so higher-status than smaller animals. Other cues, such as vocalizations, threat gestures, and the individuals' alliances and coalition partners also relate to status. Most individuals know their place in a hierarchy and thus do not fight. Fights typically occur between individuals of similar social status, those with the most to gain by rising in the hierarchy (and the most to lose if they lose their status). Hierarchies tend to be linear (A>B>C), but there are other non-linear hierarchies in which an individual's status is dependent on the support of others, or where dominance results from being in a partnership (for example, in a mated pair).

BADGES OF HONOR

Some birds have a badge of status reflecting their dominance. This is a physical attribute, such as a patch of colored feathers or beak where the more intense the color, the higher the status. For example, with white-crowned sparrows, the greater the amount of white on the crown, the higher the status. These badges need to be able to change with relative ease given changes in status, so either they are under hormonal control or will change with better nutrition. More dominant individuals may have a better diet because they have access to better foods compared to lower-status individuals. A better diet may result in brighter feathers or beak.

Right A group of hens presided over by a cockerel. Each chick knows its place in the pecking order after a series of squabbles with others over food. Winners get the food, whereas losers have to wait their turn to get the scraps.

Respecting the pecking order

Species with flexible hierarchies, with a need to recognize individuals and remember them over long time periods, inevitably will meet other groups (or large flocks will contain smaller groups, each with their own hierarchies). They need a mechanism for determining the status of unknowns based on observing their interactions and inferring their relative rank.

Individuals in different groups of pinyon jays compete over the same resources. It would seem prudent for jays to have a means of determining relative dominance, in order to reduce any potential physical aggression. As group size increases, the pressure on social memory—remembering others' relationships and dominance—increases. Therefore, it is beneficial to make judgments on relationships through observing interactions and then extrapolating to unknown relationships via transitive inference; "I know my status compared to X and Y, and I know that A is more dominant to X and Y, so I can infer the relative status of A to me, even though we have never met."

In the lab, three artificial groups of pinyon jays were formed: Group 1 (A–F), Group 2 (1–6), and Group 3 (P–S). Each group formed a linear hierarchy within and across groups. The relative dominance status of the groups was determined through competitions over food between selected members of each group (those at the highest point in the hierarchy [the most dominant] and the lowest point [the most subordinate]). Dominant birds secured the food first,

Below Pinyon jays are exceptionally social North American corvids, living in groups of dozens of individuals. Individuals form exclusive pair bonds that are supported by mutual preening and providing help when their partner is attacked by others. This form of social intelligence was recently thought to only be seen in primates and cetaceans.

frequently presenting threats, whereas subordinates would not get the food and would display submissive gestures (e.g. bowing or lowering the head). For cross-group dominance hierarchies, a limited set of individuals with a similar rank within their own groups (such as B versus 2 or A versus B) were given food competition tests, and the outcomes were observed by test birds who could use this information to construct dominance hierarchies that would inform how they should behave when meeting these unknown individuals. The birds chosen were mid-ranking, because it was important that they could both win or lose encounters (so that assessment of their status was not made using simple labels such as "that guy is always a loser").

In the experimental stage, bird 3 observed within group encounters (for example, A versus B, where B loses) and between group encounters (for example, B versus 2, where B wins). Then in the test stage, as 3 is subordinate to 2, and 2 lost to B, when B meets 3,

3 should display submissive gestures. Alternatively, in the control stage, 3 witnessed within group encounters only, which provided no useful information for encounters between groups. So 3 would see A meeting B (where B loses) or B meeting C (where B wins), but when 3 got to interact with B for the first time in the test stage, it had not been provided with information about the relative status of B to members of group 1, so 3 would not know initially how to respond to B. Pinyon jays used the information they acquired when observing between-group encounters to produce appropriate gestures when interacting with members of group 1 (submissive) and group 3 (dominant). These gestures occurred most frequently in the first minute of an encounter and then tailed off rapidly. They did not know how to respond when they hadn't been provided with information, suggesting that they weren't just responding to other physical cues. From this we can deduce that pinyon jays use transitive reasoning to infer dominance relationships.

Pinyon Jay Social Inference

An experiment to demonstrate the use of transitive inference in a social setting by pinyon jays. The text provides details of the different stages of the experiment (social hierarchy, experimental stage, control stage, and test stage).

EXPERIMENTAL STAGE

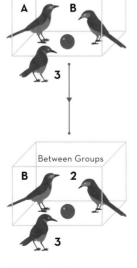

A dominant to B
B dominant to 2

CONTROL STAGE

A dominant to B
B dominant to C

TEST STAGE

after experimental stage (information about between group hierarchy)

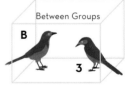

3 should be submissive when paired with B (higher status)

after control stage (no information about between group hierachy)

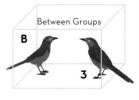

How should 3 respond to B with no information about relative status?

Standing out from the crowd

Try to imagine, if you can, that you are a young, recently hatched penguin chick living in a flock of a million Emperor penguins. The noise is deafening; so loud that it seems impossible to discern a single cry amongst the din. Now imagine that you have become separated from your parents. In such a place, how could you possibly find them again?

ONE IN A MILLION

We know that this scenario happens to penguin chicks every day during the period after hatching—so how do chicks find their family? Surprisingly, the chicks locate their parents from unique attributes of their contact call that identify them as the chick's parents, a call the chicks learn in a mere five weeks after hatching. The chicks don't even need to use all of the contact call for recognition, only the first 0.23 sec (half a syllable) and the first three harmonics (low frequencies) is enough to discriminate their parent's voices. As penguins do not create nest sites, they carry around first the egg and then their chick on their feet, so individual recognition rather than nest-site location is the only method the chicks can use to find their parents.

Other species use vocal cues to distinguish between different identities, such as jackdaws, where subtle differences between the contact call of one bird and that of another are sufficient to enable them to discriminate between birds based on the structural pattern of the call. In birds, it is primarily the contact call that is adaptable enough to be used to discriminate between individuals in this way. For other calls, such as food calls or alarm

Above Different breeds of chickens look quite distinct, with different colored plumage, exaggerated feathers, and brightly colored combs. Although these breeds have been artificially produced, we assume that each breed recognizes the others as different, and that even though we might have trouble telling individuals apart, they don't!

calls, it doesn't really make any difference who is giving the call, because invariably there will be either food or a predator at the end of it; birds don't choose to evade a predator only if bird X rather than bird Y produces the call.

INDIVIDUAL RECOGNITION

For societies in which individuals form selective, affiliative relationships with others, as well as dominance hierarchies, it is essential to be able to discriminate and differentiate those individuals from one another and remember their features over long periods. How this individual recognition occurs is largely dependent on the primary channels of communication adopted by the species in question. In the case of birds, the two main channels are sight and sound. Individuals are recognized by their unique facial features (position of the eyes, length of the beak, facial coloration, feather patterning), their individual movement patterns, their behavior, and their vocalizations. A bird that can recognize an individual should be able to do so even if the stimulus is

presented from different novel viewpoints, as it ought to be able to form a mental image of all the attributes of that individual from all viewpoints. They should be able to recognize an individual if viewed in 2D rather than 3D and ought to be able to discriminate individuals using multiple features, such as their face, posture, and voice. There are some cross-modal studies in which a subject is presented with a visual representation of a familiar bird and then played either that same bird producing a call or a different bird (or played an individual call and presented visual images of the same or a different bird). If the subject recognizes that the face and voice match, it will have expected such an outcome and so does not respond. However, if the face and voice are incongruent (that is, they don't match), its recognition of the discrepancy will be apparent in its looking behavior or vocalizations. Unfortunately, most studies are not this rigorous and simply train on a given set of stimuli, such as faces, before presenting a novel face and asking the subjects whether they have seen the new face before. This is discrimination—essential for individual recognition—but it isn't recognition *per se*.

SCANNING CROWS SCANNING FACES

Pigeons can be trained to discriminate between different human faces and even between genders and expressions of emotion. But pigeons can be trained to do many things without understanding how to use what they have been trained to do in novel situations.

I NEVER FORGET A FACE

Human faces are a category of visual stimulus, and being able to tell the difference between different faces does not mean that pigeons recognize their subjects as individuals. Pigeons process local visual information, such as the individual features of a face, whereas humans process global visual information, thus seeing the bigger picture. With faces we see the whole thing not individual components. This makes recognition of complex objects (especially from different angles) quite difficult for pigeons, but easy for us. If pigeons were required to pick out a different view of the face than the one they had been trained on, they would have difficulty because the local information would be different from what they had learned.

However, some birds are capable of discriminating between different human faces with little training and remembering them years afterwards. Birds that have to live side by side with humans as we increasingly encroach into their worlds seem good candidates for studies of human face recognition. In an experiment, urban mockingbirds (a small songbird) incubating eggs in their nests were disturbed by humans wearing masks. Across four days, the threat presented by the mask-wearing human increased as he or she moved closer to the nest. In response to the threat, the brooding female produced alarm calls and attempted to mob the intruder. On the fifth day, a new human wearing a different mask approached the nest (at the same distance as the approach on the first day). The mockingbird's behavior adjusted to that of the first day it encountered the human wearing the other mask, suggesting it could tell the difference between the two masks, although discrimination rather than recognition might explain the difference. Some birds remember human faces for a very long time, if they are memorable enough. American crows were trapped by humans wearing a mask, and the crows rapidly associated the

act of being captured with the mask. To express their frustration, the crows produced vocalizations used to mob and scold predators. They did not produce these vocalizations before being trapped (while the human wore the mask) or afterward, once they had been trapped. Therefore, they had only associated the mask with the aversive event. If the mask-wearing individual approached them even up to three years later, the previously captured crows produced the same scolding call, suggesting they had remembered their jailer.

IMAGING BRAINS

Crows recognize human faces and remember them for up to three years, and many birds are able to recognize individual conspecifics by their visual appearance, yet we know very little about the neural systems involved. The social brain systems responsible for perceiving faces are likely to be parts of a circuit including the entopallium, mesopallium, and nidopallium. Crows were captured by an experimenter wearing a "threatening" mask, but when captured were fed by an experimenter wearing a "caring" mask. The birds were kept captive for four weeks. On the test day, they were presented with an experimenter wearing either the threatening mask or the caring mask or nothing, and were then anesthetized and given a PET (positron emission tomography) scan in order to determine the nature of their brain activity at the time of the event they experienced immediately before they were anesthetized (threat, care, or nothing). By comparing their brains across the threat, care or nothing conditions, researchers were able to determine the circuits that were active in processing faces and whether there were any differences based on emotional state (negative: threat versus positive: caring). A significant increase in activity in the nidopallium and mesopallium, arcopallium, nucleus taenia of the amygdala, dorsal thalamic

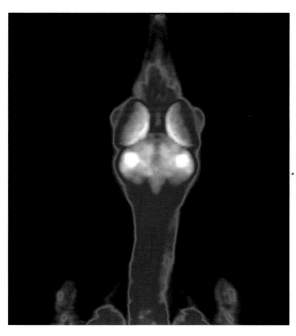

Above An American crow in a PET scanner. Its brain will remain active during the PET scan in response to what it viewed before being anaesthetized, such as the caring or threatening mask.

Right American crows have encroached far into our urban environment, and have adapted their behavior to cope with the new challenges presented by living in such so close proximity to humans, such as recognizing human facial expressions and the direction of their eye gaze.

nuclei, and brainstem was found in relation to the threatening mask, while increased activity in the hyperpallium, mesopallium, preoptic area, and medial striatum was found in relation to the caring mask. These responses are perhaps not surprising if we believe that the crow brain processes social and emotional information like the primate brain. The threatening mask elicited negative emotions via the SBN, whereas the caring mask had been repeatedly paired with reward (food) and thus activated the MRS.

NO BIRD IS AN ISLAND

When trying to survive in a social world, it pays to make friends, recognize your enemies, and determine who are the friends of your enemies. These forms of social computations require the ability to recognize individuals, but also relationships between individuals, so-called third-party relationships.

MACHIAVELLIAN MANEUVERINGS

Being able to distinguish friend from foe provides an advantage in the political arena. Individuals who cannot rely on their own wits or strength to get what they want need to foster the right connections and break the bonds between others who may thwart them in their aims. There is good evidence that monkeys and apes use social mind games to get what they want and as such they are good models for the evolution of our own Machiavellian maneuvers.

SOCIAL SABOTAGE

A variety of birds display quite sophisticated forms of primate-like social intelligence. I will focus on ravens and graylag geese because, despite being only distantly related, they have evolved similar methods for surviving in a dangerous social world. Both species form selective, long-term pair-bonds, but raven subadults may delay doing so for up to ten years. Young ravens travel in small groups, with individuals moving from one flock to another. Although not committed to one partner during this period, they form valuable relationships with others—often kin but also non-kin—that are nurtured for mutual benefit. Being in a pair, rather than having to fend for yourself, provides great advantages. Paired birds are more dominant, which helps them gain more resources—two heads are definitely better than one. The pair-bond develops when two individuals are synchronized, two social brains working in concert. Floods of the bonding hormones, mesotocin and vasopressin, through the avian emotional and reward systems reinforce the pair-bond, which is reflected in the amount of time they spend together, preening, sharing food, and supporting one another in fights. Ravens are capable of recognizing a competitor's relationships—who is dominant, who is affiliated—and of intervening in the formation of new relationships, especially if these could offer a threat to the pair's own dominance.

POWER PLAYS

Ravens recognize the dominance of others in the nature of their relationships and infer the outcome of their own aggressive interactions with an unknown assailant through their aggressive interactions with known others. Ravens recognize who is dominant to whom and will be alarmed if typical dominance relationships are violated. If ravens are played the self-aggrandizing call of a dominant, and the correct submissive response of a subordinate is played based on known dominance relations, they will pay no heed. However, if they hear a vocal interaction in which a subordinate produces a dominant call and a dominant produces a submissive call, they take notice and display high levels of stress and self-directed behavior. Like other corvids, ravens have a long social memory, each recognizing individuals and its relationship with them for years. Ravens will support their valuable partners in fights and also those that have helped them in the past.

GEESE GANGING UP

Geese also form pair-bonds, but their social groups revolve around a family unit. Geese can recognize individuals over long periods, up to at least a year, and like corvids display clear recognition of their dominance relationships. Geese pairs form strong social bonds but have been maligned in the past because the structure of their bonds is fundamentally different from those of corvids and primates. Whereas most corvids base their relationships on a mixture of tactile and non-tactile behaviors, geese restrict their bonds to non-tactile behaviors only, such as the triumph ceremony, greeting ceremonies between allies, vocalizations, social support, behavioral synchronization, and spending time in close proximity. Corvid displays are definitely more primate-like, but not necessarily more complex and not necessarily indicative of a stronger bond. Like corvids, members of

Above Like most corvids, raven pairs spend the majority of their time together, even in flight. They keep in touch by producing a contact call that has an individual signature, so their partner can quickly find them if they become separated.

a goose pair are more successful in gaining resources by increasing their dominance compared to singletons, and members of the pair will also support one another in a fight (active support) and be positively effected by the presence of their family, who may not take part in fights (passive support); this presence in turn reduces physiological responses to stress and leads to enhanced health and well-being.

Right A flock of greylag geese foraging. Although they form large social groups, the goose social system is primarily at the level of families, with a bonded pair and their offspring. Once the goslings have hatched, they imprint on their mother and follow her wherever she goes, leading them to places to feed.

The cooperative bird

Many animals help one another, but typically for selfish reasons. Mutualism is a form of cooperative behavior in which two agents gain benefits by aiding each other. Parental behavior—in which two adults cooperate to build a nest, find food, and feed chicks that will aid them in producing viable young—is an example.

KEEPING IT IN THE FAMILY

Cooperative behavior in birds does not always seem beneficial. Florida scrub jays and Arabian babblers are cooperative-breeding species, in which only the dominant pair get to mate. Other members of the family, usually the offspring of the dominant pair from a previous year, will help out rather than breed themselves. Helping out consists of looking out for predators and helping to mob them if they get too close for comfort. Arabian babbler sentinels will produce

Above Rooks, like most corvids, form life-long relationships with one partner. To cement this bond, partners' use a form of physical contact called bill-twining, resembling human kissing, where they either nuzzle each other's bill or hold their partner's bill in theirs.

alarm calls as soon as they see a predator threatening their new brothers and sisters. Young Florida scrub jays are called helpers at the nest as they also assist in feeding new chicks. Why do they do this and not find their own mate and produce their own young? With regard to passing on genetic material to the next generation, the percentage of genes shared between two siblings (brothers and sisters) is the same as that shared between parents and their offspring. When environmental conditions are tough—perhaps it is a bad year for food resources, so not enough is available to sustain a larger population—then it makes more genetic sense to help raise your siblings, and increase their chances of survival, than to raise your own brood.

I'LL PREEN YOUR FEATHERS, IF YOU'LL PREEN MINE

Although it is assumed that cognition does not play a role in mutualism, other more complex forms of cooperation, such as reciprocity, may require different cognitive abilities in order to function successfully. By offering a resource now for a promised (different) resource in the future, you are holding off receiving an immediate reward (temporal discounting). If I preen you now, I will expect to get something (of equal value) in return later. I therefore have to discount this future commodity until I receive recompense later. Another ability that is useful to possess in a cooperative species is some form of numerical discrimination. This is important to make sure that commodities of equal value are being exchanged. Social support, for example, may be valued much more highly than the delivery of a morsel of food because of the potential physical risks involved in taking part in a fight. Therefore, three or four food shares may be seen as equivalent to one act of social support. A final essential skill is that of long-term memory. If you provide assistance now, you may not require recompense until later, some time in the future. Then, you will need to remember that you are owed a debt that has to be repaid. Again, there is evidence that many birds have long memories lasting years for food and faces.

PULLING TOGETHER

Despite many studies of cooperative behavior in wild birds, the experimental study of avian cooperation is not very advanced. One task that has been used to examine what primates understand about when to cooperate and when to work alone, as well as who to cooperate with and when, is the so-called cooperative string-pulling paradigm. Subjects are presented with a wooden platform with two hooks screwed into it, through which a long piece of string is threaded so that pieces of equal length come out of each end. Resting on the board are one or two food bowls containing tasty treats. As the board is out of reach, the only way to get the food is to pull on both ends of the string at the same time. Pulling on only one end will cause the string to pull through the hooks. For one bird to pull on both ends of the string at the same time, it first needs to bring them together and then pull. If the string is too short to bring the pieces together, then you need two birds and each has to pull its own end of the string at the same time as the others. Rooks presented with this task learned how to pull the strings together to move the food toward them. They also worked as a team pulling in the board when the strings were too short to pull in alone. However, they only did this with birds they tolerated standing next to and feeding with. When the rooks had to delay pulling until a partner had entered the room, they could not (consistently) wait. So rooks can work together as a team but cannot always curtail their "excitement" when seeing the food so as to wait for a helper to solve the task. A similar study in African grey parrots found the same response.

Rook Cooperation

Illustration of rooks performing the cooperative string-pulling paradigm.

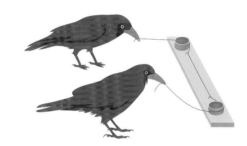

1 The string is long enough for the rook to bring the two ends together to pull on them with assistance.

2 The string is too short to be able to pull in alone, so rooks work as a team.

3 The rook has to wait for a partner to arrive before pulling, but fails to do so.

Repairing broken relationships

One cost of living with others is that it requires the sharing of resources. Although dominance hierarchies make aggression less likely, fights do happen. This is to be expected between rivals, but fights also occur between friends, family members, and partners.

WE ALL NEED FRIENDS

The best way to survive a complex social world is to make friends and minimize the number of your enemies. However, to make friends you need to be able to share. Without the fair sharing of resources, disagreements occur, but as disagreements among friends are not so damaging as those between enemies, there are ways that a broken relationship can be salvaged and even repaired. This is called reconciliation.

STRESS MANAGEMENT

The concept of reconciliation was first described for primates. In the aftermath of a fight between two protagonists, they are most likely to move as far apart as possible in order to reduce the chance that aggression with reoccur. What is going on in those cases when the protagonists don't separate but rather come together and affiliate? In some extreme cases, such as chimpanzees and bonobos, one-time rivals will kiss and even copulate. Reconciliation between the former opponents is an attempt to mend a broken relationship. This doesn't happen after every fight and doesn't happen between all former opponents, only between those with a strong enough relationship it is more important to save than to lose; perhaps they are kin or perhaps they are alliance partners who cannot survive alone. Reconciliation has now been described for a number of mammals, including primates, carnivores, dolphins, and whales, and even some sheep and goats.

What about birds? Most birds are monogamous, so the pair-bond represents the best case for a

valuable relationship. However, in birds that form lifelong pairs, such as rooks, there is no evidence for reconciliation because there is no aggression between partners. As they don't fight, there is nothing to reconcile. Subadult ravens, before forming pair-bonds, have other valuable relationships, primarily with kin. Ravens who fought a partner with whom they had formed a valuable relationship were more likely to approach and then affiliate with that partner after a fight than those who did not have a valuable relationship with their opponent. The likelihood that a fight would occur after reconciliation between these former opponents was vastly reduced compared with individuals who had not been reconciled. Reconciliation was also more likely to occur when the fight was of greater intensity.

The PC-MC (post-conflict : matched-control) method is used to measure reconciliation. Ravens were observed interacting in a social group, but once an aggressive encounter had happened, the frequency of any affiliative behavior that occurred between the two fighters was recorded for a period of ten minutes after the fight ended. This is called the post-conflict period. To determine the baseline level of affiliation between the two birds in their normal day-to-day interactions, affiliative behavior was recorded during a comparable ten-minute period at the same time on the following day, in this instance without any preceding aggression. These two periods were compared, with the expectation being that affiliation would be higher in the PC than in the MC period for the first few minutes after aggression; this was indeed found to be the case for ravens.

Left A raven preens its partner. This important behavior serves both a cleaning role (removing parasites), and a social function (solidifying a long-term bond). These are the same functions as grooming between closely bonded monkeys or apes.

Raven Reconciliation

An aggressive raven (A) approaches another raven (B) and attacks. A is more likely to approach B after the fight and preen it if they have previously formed a valuable relationship.

Aggressor (A)

Victim (B)

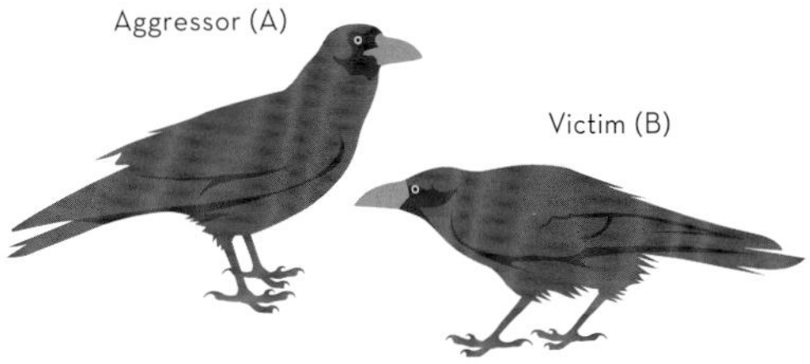

1 A approaches B

2 A attacks B

3 A preens B (reconciliation)

LEARNING FROM OTHERS

A major benefit of living with others is the exchange of information. Rather than learning everything first-hand, which can be dangerous, or having every skill inborn, which is rather inflexible, being able to learn from others can be highly advantageous. Learning from others' mistakes vastly increases your chances of survival. Birds learn socially when, where, what (and what not), and how to eat, but also who might make the best mating partner and what the best method is of evading predators.

SOCIAL INFLUENCE

Social learning is a broad umbrella covering many processes. The simplest forms of social learning are social influences, such as behavioral contagions in which we automatically respond to others behavior. It is difficult not to laugh, cry, cough, or yawn when another person does the same. This is why when one dog barks the whole canine neighborhood joins the chorus. Just being in another's presence is sufficient to influence your behavior. You may not be hungry, but seeing someone eating can be a sufficient trigger to stimulate hunger. Social presence can increase your arousal and motivation, affecting your behavior. This can be most acute when learning about which foods to eat when presented with something new. Rooks, for example, are more likely to eat a new food that a neighbor is eating than another novel food the neighbor isn't eating. Many birds will eat a greater variety of foods in another's presence than when alone, even though the potential negative effects of making a wrong choice—such as illness—do not often occur until some time after consumption.

ENHANCEMENT

An especially powerful form of social learning is enhancement. This draws another's attention to a location, object or event that is of specific interest. In local enhancement, an individual's attention is directed to a particular location, perhaps an especially rich patch of tasty food. This behavior does not need to be intentional, so an individual doesn't need actively to direct another. Rather, the individual's presence in one location is sufficient to elicit interest. A flock of birds forages on a patch of farmland and this attracts other birds to the same patch. They have become attracted to this spot because of the other birds feeding—a significant sign that food can be found there; it is more efficient to follow this lead than to go looking for food elsewhere, which requires a significant amount of energy.

In stimulus enhancement, an observer's attention is drawn to a specific object or event, independent of its location in space. The most famous example is of blue tits opening the foil lids of milk bottles in the UK in the 1940s. At the time, milk bottles were delivered and left on doorsteps. The cream would float to the surface of the milk and provide a tasty treat for birds. Blue tits would observe other tits drinking the cream after breaking through the lids. This behavior spread quickly throughout the UK and was presented as an example of culture. However, lab experiments found that the blue tits were not copying the precise actions of a model, rather they were attracted to the tops of milk bottles, learning through trial and error how to get the tasty cream inside. Therefore, nothing more sophisticated than stimulus enhancement was at play in spreading this technique.

LEARNING TO BE SCARED

Another type of social learning is observational conditioning. A previously uninteresting object is made aversive or attractive by its association with another's positive or negative reactions to that object. I developed arachnophobia through seeing my mother frightened of spiders, even though I did not have any bad experiences of spiders personally. Blackbirds can be conditioned to fear non-threatening objects, such as bottles or honeyeaters, when they observe a conspecific being aggressive toward a little owl (a predator that only they can see, while the observer can only see the non-aversive object). The observer therefore associates the negative reactions of the other blackbird with the neutral object, thereby forming an aversion to that object.

Above When British households used to get their milk delivered to their doorsteps, blue tits were renowned for their love of the cream floating on the surface of the milk. Blue tits quickly learned to peck the silver lids of the milk bottles to get at the delicious cream underneath, but it's unlikely they learned this through imitation.

EMULATION NOT IMITATION

A final example of non-imitative social learning is emulation. An observer learns about another's goals through watching their actions, but does not copy the method used to achieve those goals. A bird is given a puzzle box in which a number of devices need to be released in order to gain access to the food inside. Some devices are distractors, playing no part in reaching the food. However, if subjects observe a demonstration of how to open the box that includes manipulating these distractor devices, they may either faithfully copy the precise actions they observed (imitation) or complete the task using the most proficient and efficient method possible (emulation). The goal is the same, but the method for achieving the objective is different. Emulation is the smarter strategy. Indeed, in discussing imitation next, we will see that birds with smaller brains rely on imitation as a strategy to learn about the world.

Right A small foraging flock of cockatoos in a Canberra park. Such social birds learn from members of the group where to look for the best sources of food, what might be good to eat and what should be avoided, and even how to get into hard-cased foods, such as fruits and nuts.

Watch and learn

Imitation involves the precise copying of another's actions, but that doesn't mean that the observer has to understand why the demonstrator is performing those actions.

DO AS I DO

A problem with ascribing social learning to imitation is that there are often alternative explanations, such as stimulus enhancement. For example, a bird plays with a tube with food in the center, with cotton wool stuffed in both ends. The simplest way to get at the food is to remove the cotton wool from one end. For most birds, the only way to achieve this is to pull the wool out with the beak. If an observer performs actions that match those of a model, what social learning mechanism was at play? We would be correct in attributing the observer's actions to imitation rather than stimulus enhancement, because it copied the demonstrator's actions closely. However, due to the nature of this task, there wasn't an alternative action it could have used to remove the wool.

THE TWO-ACTION PROCEDURE

A method that has been used to avoid this problem is the two-action procedure. A task is devised in which one of two different actions could be used to reach a goal. Half the subjects see a demonstration of one method, while the other half sees a second method. When they are provided with the apparatus, the subjects are observed to see whether they match the method they observed. We expect birds that can imitate to copy the method they witnessed precisely. In the case above, half the birds would see cotton wool

Below Japanese quail chicks are precocial, so fully formed at hatching. Being short lived, they have to deal with the dangers of the world quickly, rather than learn about them through personal experience. This may be a reason why quail are capable of learning by imitation, whereas longer lived, altricial birds, such as crows, cannot.

removed by pulling it out with the beak, while the other half would see it removed by shaking the tube. In this case, stimulus and local enhancement or emulation aren't sufficient alternative explanations, because both actions were directed to the same object. Stimulus enhancement or emulation would not explain the observer's performance because it copied the exact method it saw being performed. Some of the clearest cases of imitation come from work in precocial birds (those that are fully developed when hatched) rather than from primates. Because these birds have relatively small brains, this suggests that imitation as a strategy may not be especially intellectually taxing or require flexible thinking.

PECK OR STEP

One of the best examples of imitation using the two-action test is a study of Japanese quail. Demonstrators were trained to perform one of two actions to operate a treadle in a test box that dispensed food. They either pecked the treadle or stepped on it—both actions causing food to be released from a feeder. Once the demonstrators had been trained, an observer was placed into an adjacent chamber with a window between the two birds, so the observer could witness a demonstration of one of the techniques—half the subjects saw the peck method and half saw the step method. Because the demonstrator's actions were directed to the same object (that is, the treadle), this eliminated explanations based on stimulus enhancement. When placed in the test chamber, those quail that had witnessed the peck technique tended to use the peck technique to release food, whereas those that had observed the step technique tended to use stepping to release the food. This result suggests that the birds imitated the method they had previously seen and added it to their own behavioral repertoire. For a bird with a little brain, this may be the best method for learning about actions that would otherwise be outside their individual capabilities without extensive training.

Quail Imitation

Illustration of the two-step procedure in Japanese quail. Left: The demonstrator bird performs pecking on the treadle while the observer watches. Right: The demonstrator bird performs stepping on the treadle while the observer watches.

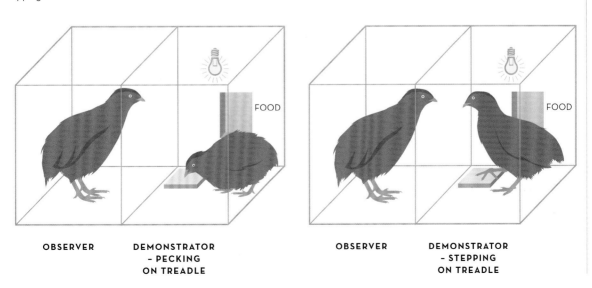

| OBSERVER | DEMONSTRATOR
– PECKING
ON TREADLE | OBSERVER | DEMONSTRATOR
– STEPPING
ON TREADLE |

5 THE RIGHT TOOL FOR THE JOB

An extension of beak and claw

Until primatologist Jane Goodall found that chimpanzees fashion sticks to probe for termites, humans were thought to be the only toolmakers. Hundreds of examples have since been discovered in the animal kingdom—the most numerous tool users being birds and mammals. Yet tool use remains rare and toolmaking even rarer.

TOOL USING AND MAKING

For our purposes, tool use is the use of an unattached, external object to increase the efficiency of a user's body to achieve a goal or state not possible without the use of the tool. Therefore, using a stick to gather an out-of-reach fruit represents an example of tool use. A beak isn't sufficient to reach the fruit alone, so a bird needs a stick to extend the reach of its beak or bill.

Most species do not need to alter an object for it to function as a tool. For example, a stone can be used to break open a shell without having to sculpt it. Some species do change objects to make them function as a tool or to work more efficiently. Twigs attached to trees replete with leaves are tools in their raw state and without modification would make poor tools for inserting into holes to extract insects. The most sophisticated tool crafting is found in humans. Think of the stone hand axes of our earliest ancestors.

WHICH BIRDS USE TOOLS?

Tool-using birds are found in two main groups: songbirds (including corvids) and parrots. These groups contain species with the largest relative brain size and possess intelligence equivalent to the great apes. But there are also exceptions of tool-using birds not found in these groups. Egyptian vultures, for instance, aside from scavenging, forage on delicacies such as ostrich eggs. The vultures' problem is how to get into the eggs, because the shells are extremely tough. The eggs are too heavy and awkward to lift, so instead of dropping them onto a hard surface, as with thrushes hitting snails onto pavements, Egyptian vultures have learned to drop stones onto the eggs to break them open so that they can feast on the contents. In terms of definitions of tool use, using rocks to break open shells is classified as such, whereas dropping shells onto a hard surface is not.

Left An Egyptian vulture attempts to break open an ostrich egg by dropping rocks onto it. This is a simple form of tool use.

Another classic example of avian tool use is heron bait-fishing. A heron will grasp small insects in the tip of its bill and hold them over the surface of the water. Fish attracted to the insects rise to the surface, becoming targets. Burrowing owls do something similar, placing dung at the entrance to their burrows to attract insects, such as beetles, which are then eaten. Both the insects and the dung could therefore be classified as tool use.

Some birds only use and make tools in captivity. One example is a Goffin's cockatoo, Figaro, who lost a stone he was playing with outside his aviary. Researchers observed him using a piece of bamboo as a tool to (unsuccessfully) reach the stone, so they tested his capability to make tools to retrieve out-of-reach food. They found he made a series of tools to reach the food by stripping off pieces of wood from his aviary. This species does not use tools in the wild and this type of toolmaking is not a trivial task for a curved beak. Similar forms of tool innovations have been reported in captive blue jays, marsh tits, hyacinth macaws, and rooks.

What might be classified as tool use is the behavior of urban crows in Japan and California. Crows enjoy feasting on nuts, but it is difficult to crack open the hard shells, so the crows have devised a strategy to do this. Dropping the nuts onto the road, they waited for a car to drive over them, crack them open and disperse the contents. However, this of course leads to an additional problem: how to retrieve the goodies without getting run over? The crows have gotten around this by dropping the nuts onto a pedestrian crossing, waiting until the crossing turns green and collecting their treats in safety! But by way of a caveat, studies by California scientists suggest that some crows may not be using cars as tools. They studied the frequency with which crows dropped nuts when cars either were or were not present. They found the birds were as likely to drop nuts when cars were present as when not, suggesting that their actions were not directed towards the goal of cracking open the nuts.

A world distribution of avian tool use

American Crow
Probe Tool

Green Heron
Bait Fishing

Blue Jay
Reaching Tool

Marsh Tit
Stickers to Store Food

Woodpecker Finch
Probe Tool

Hyacinth Macaw
Wedge Tool

Burrowing Owl
Dung as Bait

Of the 10,000 species of birds, only a few are known to use tools and, of those, only a handful have been observed making tools in captivity. Only two species, woodpecker finches and New Caledonian crows, habitually make tools in the wild. Tool use and making is therefore extremely rare in the avian world.

Observation in the wild

Captivity or lab experiment

Tool user and manufacturer

Tool user only

Rook
Hooks, Stones, Sticks

Egyptian Vulture
Dropping Rocks

African Grey
Scratching Tools; Cups

Goffin's Cockatoo
Probe Tool

Jungle Crow
Dropping Nuts on Roads

New Caledonian Crow
Probe Tools

Satin Bowerbird
Paint Bower Ornaments

Kea
Probe Tool

Why are there so few tool-using birds?

Why do so few birds use and make tools? If tools were so useful, wouldn't we expect them to play a more important role in the lives of more species? Certainly, using a tool to procure their next meal would advantage many animals. Predatory species possess the anatomical specializations that make them efficient killers and don't need tools to secure their prey, whereas most species that hunt for grubs hidden inside holes and under tree bark have to employ tools, because they don't have the powerful claws or jaws that could rip apart the trees concealing the insects.

IS TOOL USE CLEVER?

Across birds and primates, those species with bigger brains and specifically a larger cortex (or nidopallium) are more likely to use tools than those with a smaller cortex. However, those same species also live in large social groups, live longer, and demonstrate problem-solving not associated with tool use. Therefore, clever species tend to do lots of clever things and generalize their intelligence to the solution of a range of problems. Can we say that tool users are smarter than non-tool users? We would be right to assume that using a tool requires a level of brainpower over and above that required for a similar process that doesn't employ a tool, because of the additional processing required with the extra object. Tool users have to think about their tool's form in order to increase its efficiency. When attempting to reach food, a stick with a hook or bend at the end is going to make a more practical tool than one with a straight end. A tool user may select a stick with a natural bend, whereas a toolmaker would go a step further and create a bend at the end of the stick. We might suggest that a toolmaker requires more brainpower than a non-tool user. Indeed, the rarity of non-human toolmaking species suggests that this should be the case. But what evidence exists that tool users are smarter than non-tool users?

Studies on the cognitive abilities of toolmaking species such as woodpecker finches and New Caledonian crows have found little evidence to support this claim. Crows, for example, have a rudimentary understanding of the relationship between cause and effect and also understand when a particular task needs a tool with a particular length, diameter, or flexibility. However, other studies suggest their causal reasoning is limited and they cannot make simple manipulations of a tool, such as flipping it into the correct orientation when used for a different task.

Above A New Caledonian crow using a hook stick tool to remove a grub from a hole in a fallen log.

HOW DID TOOL USE EVOLVE?

Perhaps a stronger argument against the idea that there is something special about the intelligence of tool users is that the performance of non-tool-using birds, such as rooks and small tree finches, is equivalent to that of the tool-using birds. Does this tell us anything about how tool use evolved? If there is no cognitive advantage to being a tool user and non-tool users understand the physical properties of objects such as tools, then why are there so few tool-using species? Many species may have the brains to use a tool, but not the opportunity in their particular environment, so that tool use may not be as ubiquitous as we might think it should be because most animals' diets are perfectly adequate without needing to resort to a tool. Chimpanzees, for example, use tools for a number of their food-acquisition and -processing needs, yet most of their dietary requirements are fulfilled without tool use. Indeed, the highest calorific food eaten by chimps, colobus monkeys, is acquired using cooperation, not tools.

THE RIGHT TIME AND PLACE

Many birds, especially corvids and parrots, are innovative—they come up with solutions to novel problems—and their levels of innovation correlate with their brain size. They also extensively manipulate their environments, playing with objects, including those that could be made into tools. These species should have the opportunity to make tools or use objects as innovative solutions to novel problems. So why don't they? We have ruled out the possibility that they don't have the brains (at least for tool use) or the necessity. What about their environment? Common features of the habitats of tool-using species are a lack of competitors for the foods they would acquire with tools and a lack of predators, so that they can afford the time required to make an efficient tool. Woodpeckers eat grubs, but woodpeckers are not found on New Caledonia or the Galapagos Islands; neither location has any endemic reptilian or mammalian predators. By contrast, rooks would have to compete with woodpeckers and there are also many predators to distract them from creating tools. Finally, rooks eat vegetable matter and insects found under the ground that can be reached by the use of their thin, straight bill alone. No tool required!

Right A Goffin's cockatoo, Figaro, sees a cashew out of reach, strips a piece of wood from his aviary to use as a tool, pokes it through the mesh, pulling the nut towards him and finally bringing it close enough to eat.

DARWIN'S TOOL USERS

When Charles Darwin traveled to the Galapagos Islands in 1835, he collected a number of species of finch whose patterns of behavior eventually formed the basis for his theory of evolution by natural selection. One Galapagos finch not seen or collected by Darwin is notorious for its tool-using abilities.

Woodpecker finches are so named because they have taken over the ecological role of woodpeckers on the Galapagos, eating grubs found under tree bark. They use two types of tool—cactus spines and twigs—which they insert into holes and under tree bark to extract the grubs hidden there. They are members of an exclusive toolmakers' club because they modify their tools, shortening them if too long or removing extraneous side branches and leaves from twigs. The Galapagos Islands is a perfect environment for tool use to evolve because there are no competitors, no predators, and the ecology and climate are unpredictable. Evolving a method capable of securing a more diverse range, a greater number, and a better quality of prey, especially in times of paucity, would be extremely beneficial to those individuals who use tools.

The study of tool use in woodpecker finches has been the result of a happy circumstance. The finches are found on the inhabited island of Santa Cruz, which has a variety of distinct habitats. The birds can be easily captured and kept temporarily in captivity to take part in cognitive experiments. This work is the result of a long-term study by Sabine Tebbich and colleagues. They have studied the prevalence of tool use in Santa Cruz across different seasons and climatic zones, as well as the effects of social learning on the development of tool use in young finches; and captive-housed finches have been given various cognitive tests to determine whether their tool using is intelligent.

FOR A RAINY DAY

Not all woodpecker finches use tools; tool use is restricted to the Arid Zone of Santa Cruz and takes place only during the dry season. In the dry season, grubs can only be found under tree bark and are inaccessible without the use of a tool. By contrast, in the wet season, grubs come to the surface and are plentiful and can be extracted without a tool. In the Arid Zone dry season, tool use accounts for 50 percent of foraging time and produces 50 percent of the birds' diet. The opposite Scalesia Zone is forested and it rains there all year round. Arthropods are in abundance in this zone throughout the year and forage easily without tools. Tool use in this region is almost negligible.

As tool use is dependent on climatic and geographical variability, do all young woodpecker finches learn to use tools, is it restricted to those individuals that will experience these harsh environments, or are all finches born with the ability? To find out, Tebbich and colleagues collected two broods of recently hatched finches and divided them

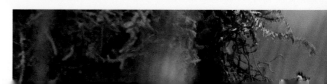

Left A woodpecker finch on the Galapagos Islands uses a thorny stick to forage for insects in a tree crevice.

into two groups; one group received demonstrations of tool use from conspecifics, while the second group received no opportunity for social learning. Both groups developed into equally proficient tool users suggesting that finches are born with a skill for fashioning and manipulating objects as tools. This makes sense when you realize that woodpecker finches are rather antisocial birds, with little opportunity to learn outside the short period they remain in their parents' care before adult independence. Such a small time window would be insufficient to learn a complex skill such as using and making tools, so being born with the ability, or at least a predisposition to learn quickly, makes sense.

LITTLE EVIDENCE OF UNDERSTANDING

Cognitive studies on woodpecker finches suggest that they are rather limited in their understanding of the physical world. They were capable of selecting the correct length of tool and were able to modify H-shaped and S-shaped tools to make them functional (i.e. fit inside a Perspex tube so that they can push food out the other end) but only when they had failed with an unmodified tool. When compared to non-tool-using small tree finches, woodpecker finches displayed no advantage on a reversal task (testing behavioral flexibility) or a seesaw task (when perching on the correct lever provided access to food and perching on the incorrect lever caused the food to fall into an inaccessible gap). Woodpecker finches were more successful at a box-opening task (opening a novel box by flipping up a transparent lid), an aptitude that may be related to their feeding ecology, which requires lots of pecking, as they would eventually learn to peck open the box lid.

Right A woodpecker finch using a twig to search for food. During the dry season it is hard to extract grubs without the use of a tool.

The master toolmaker

Of the two bird species to use tools habitually, only New Caledonian crows, found on the Pacific island of New Caledonia, sculpt tools from raw materials. Woodpecker finches only make occasional small changes to cactus spines, such as adjustments to the length of the tool, because the spines are already decent tools as nature made them, with a sharpened end that can be used to spear an insect prey. By contrast, New Caledonian crows make two types of tools that have to be fashioned into tools from their raw state.

TOOLS FROM LEAVES

The first type of tool are made from *Pandanus* leaves. *Pandanus* trees, found throughout New Caledonia, are ideal materials for making tools, as their leaves are both strong and flexible. A crow approaches an individual leaf, bites one end and then pulls off a long strip, detaching the whole piece from the leaf at the end (see diagram). The tool has a sharpened point at one or both ends, and due to the nature of the leaf has a series of barbs along one side of the tool. Crows have been found to make variations of this tool that are more or less efficient in catching particular prey.

These different types have been found across specific regions within New Caledonia. Some leaf tools are long and thin (narrow tools), and thus relatively weak and flexible, whereas others are stiffer and thicker (wide tools). Finally, the most common type are called multistep tools, which are an irregular thickness and the most at their thickest point, but also flexible at one end, aided by a series of steps along one edge of the tool. These steps make the tool stronger at one end, but thin and flexible at the other, so easier to insert into a crevice to extract a grub. You can see how these different tools are made in the following diagram.

Pandanus Tools

How different Pandanus tools are made (A–D). A crow makes a small cut in one edge of the leaf along the edge with barbs. It then makes a sufficient number of tears to create the required tool before removing it from the leaf. This suggests that the toolmaker has an image of the required tool in its head before making it.

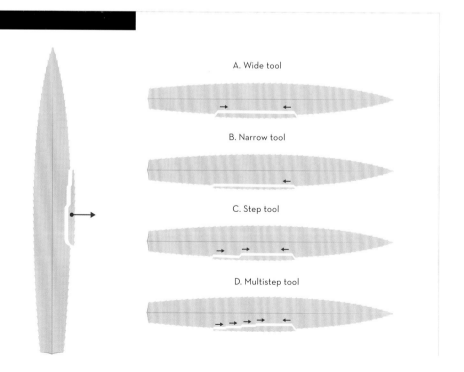

A. Wide tool

B. Narrow tool

C. Step tool

D. Multistep tool

MAP OF NEW CALEDONIA

Cumulative Culture

Map of New Caledonia with the distribution of different tool types found across the island. It has been suggested that this distribution relates to a cumulative culture; crows in different regions have adopted a specific type of tool from a simple, early form that may have spread and be evidence of culture. However, the evidence for this is weak.

KEY

Multistep tool

Wide tool

Narrow tool

TOOL CULTURE?

Once these tools have been made, they leave a counterpart on the leaf, a ghost of the departed tool. A survey of *Pandanus* leaves across New Caledonia revealed that different populations of crows used different *Pandanus* tools across the island. Only wide tools were found in the southeast corner, whereas only narrow tools were found in a larger area of the southeast. The most common tools found throughout the island were multistep tools of various types and with a different number of steps. Gavin Hunt and Russell Gray, who made this discovery, suggested that this pattern of tool distribution over the island might represent a case of cumulative cultural evolution. This is where an artifact displays obvious signs of having been tinkered with by a maker, changes that occur incrementally over a wide geographical range. The simplest form of the artifact is created in one site, such as the wide tools in the southeast corner of New Caledonia. Then an individual innovates a new design that functions better—either more efficiently or using less material or less effort to make—whereupon this individual moves to a new location, taking the new design with them. Individuals in this new region, say further into the southeast of the island, incorporate this new design into their repertoire, find that it works better, and use it exclusively so that the old design is no longer used, only the new design. This scenario can only occur through social learning (see Chapter 4). Finally, a further improvement on this design is made, through either the insight of a toolmaker or a happy accident, and the new design has been improved.

Left A New Caledonian crow uses a tool created from a twig. A natural hook was fashioned from the end that was attached to the original branch. The hooked end is inserted into a hole in the tree bark and used to fish for huge grubs living inside the tree. The hook is used to agitate the grub, which eventually grabs onto the stick and can be retrieved and eaten by the bird.

The idea of cumulative culture is intriguing and attractive, but unfortunately there is little evidence to support it. Despite the size and complexity of their social groups, there is little evidence that corvids can socially learn the sort of properties of objects that would enable this form of culture to occur. Also, the evidence of different tools across New Caledonia is dependent on finding the leaf counterparts that result from making them rather than observing the crows making the tools. Because New Caledonian crows are not especially social birds, they may also work out the design of a new tool by reverse engineering from observing counterparts on leaves. It is possible that the New Caledonian crows use a variety of different *Pandanus* leaf tools for different functions, mostly multistep tools (equivalent to Swiss army knives), with narrow and wide tools being reserved for specialized tasks. I can use my multi-tool for all sorts of jobs, but I resort to a proper screwdriver when I need a shorter or longer shaft for specific tasks. Perhaps narrow or wide tools are required in southeast New Caledonia because of a specialized task restricted to that part of the island alone.

TOOLS FROM STICKS

The other main tool created by New Caledonian crows is the hook-stick tool made from relatively long sticks broken from low-hanging trees. These tools are much more rigid than *Pandanus* leaf tools and may serve a different function. They differ because of the addition of a hook at the end that can prise out grubs in awkward locations where leaf tools simply can't penetrate. To make the hook tools, crows select a series of interconnected twigs with the leaves still connected. They systematically detach smaller twigs and leaves making the branches easily to handle, then remove the larger branch producing an upturned hook at one end. The crows continue the process of removing and sculpting until they have completed a finished natural hook-shaped tool, although they will also continue to make changes as needed. Indeed, as far as we are aware, no other animal makes as many small changes to their tools—even during use—as do New Caledonian crows fashioning hook sticks.

Stick Tools

How hook stick tools are made.

1. Select fork

2. Remove and discard all side twigs and leaves

3. Remove tool twig

UNDERSTANDING HOW TOOLS WORK

Although the question of whether a nonverbal creature understands anything is a difficult one to answer, we can at least provide them with various tasks with one or more solutions and then, using various control procedures, ask whether they solve the task and how they do so. Do they solve it immediately or after a number of attempts? If solved immediately, do they need only to look at the task to solve it or do they require some handling time? A creature may solve a task with imagination and insight, but is more likely to do so using a simpler process, such as trial-and-error learning.

There is great disagreement among comparative psychologists over what mechanisms may play a dominant role in how animals use tools. It may seem obvious that use of a tool suggests knowledge of how that tool works. Yet even after decades of study on various tool-using animals, it is far from clear that they know what they are doing when they use a tool. That is not to say that they are not successful in using the tools; rather that, when they are tested in an experimental foraging setup, their understanding of tools—assessing which tools are the most appropriate for which task and seeing the physical constraints of those tools—is not straightforward.

A FAILURE TO UNDERSTAND

Despite some animals' natural tool-using abilities, laboratory experiments often fail to find comparable performances when tested in the context of their understanding of the most appropriate tool to use or the consequences of using one tool over another.

One method comparative psychologists have used is to provide a scenario where using a tool in one context leads to the attainment of a reward, whereas in a different context it will lead to the reward being lost. A typical test for this is the trap-tube task, designed to emulate the natural foraging of primates inserting sticks into holes to extract grubs. A chimp, for example, inserts a stick into a termite mound; the termites are agitated and grab hold of the stick, which the chimp withdraws to eat its prey.

THE TRAP-TUBE TASK

In the trap-tube task, subjects are provided with a clear plastic tube with a treat inside. In the simplest version, the subject has to use a stick to poke or pull the treat out of the end of the tube. The task is made more difficult by adding a trap to the base of the tube located in or near the center next to the food. To retrieve the food, the subject has to avoid pulling it into the trap where it cannot be retrieved. New Caledonian crows and woodpecker finches have been given this task with mixed results. A New Caledonian crow, Betty, learned to solve this task in 60 trials, but was probably using a simple rule, "Avoid the side with the object under the tube [trap]," because when the tube was rotated so that the trap was now on top of the tube (and so could not function as a trap), Betty continued to use this rule. When woodpecker finches were tested with a transparent and a covered trap tube, one bird, Rosa, solved the task within 50 to 60 trials but, unlike Betty, chose to insert a stick into either side on an equal number of occasions when the trap was inverted.

Non-tool-using birds have been tested on a modified version of the trap-tube task where either a stick was pre-inserted or the bird could directly manipulate the reward without having to use a tool. Parrots (keas, macaws, and a cockatoo) were successful on tubes in which they could directly move the food, and rooks solved a trap-tube task within 30 to 50 trials when pulling a pre-inserted stick.

Left A chimpanzee has stripped a twig of its leaves, and used the twig as a tool to fish for termites. It inserts the twig into a hole in a termite nest, the termites grab hold of the twig, which the chimpanzee withdraws scooping the clinging termites into its mouth for a tasty and nutritious treat.

Right Although most birds do not use tools in the wild, there are many cases in which captive birds have been demonstrated to be capable of using objects, such as stones and sticks as tools, as well as performing successfully on tool-related tasks. This suggests that many birds have the cognitive abilities necessary for tool use.

The consequences of our actions

Tool users may work out how to use a tool from how it looks (their perceptible features) rather than understanding how it will work. However, can some animals imagine the link between their actions on a tool and the resultant effects of these actions? Such causal reasoning should aid a tool user in making the most efficient use of a tool. A classic method for testing an animal's ability to compute cause and effect is the trap-tube task, but as we have seen, there are simpler ways to solve this task.

THE TWO-TRAP-TUBE TASK

We designed a modified version called the two-trap tube task with a stick pre-inserted and an additional non-functional trap that had the appearance of a normal trap but allowed the food to be retrieved. Two configurations of this task were presented. In Tube A, the trap was reversed so the base was continuous with the tube, with the result that pulling food across the trap allowed food to be retrieved from an open end. In Tube B, the trap base was removed, so that food could be pulled into the trap and retrieved underneath the tube. These tubes were perceptibly different, but conceptually the same. If birds were tested on Tube A and they understood the concept, they should be able to solve Tube B. If they hadn't learned the concept and were basing their actions on perceptible features, then they would need to relearn every new tube. Eight rooks were presented with tubes; half received Tube A and half Tube B. Seven birds rapidly learned the original tube and immediately transferred to the other tube. When retested with the original tube, they displayed a perfect performance. These results suggested they could reason as to causes as well as tool-using animals. However, a more skeptical voice might say that they had learned a simple rule such as "avoid the trap with the dark base," because both tubes A and B had functional traps that were the same.

HOW TO SOLVE THE TWO-TRAP TUBE TASK

We therefore attempted to eliminate perceptible features or simple rules as explanations for the rooks' behavior by creating two new tubes with two non-functional traps pitted against one another. One trap had no base and the other had a raised base so that food could pass over the top. Both non-functional traps had been equally rewarded in previous incarnations, so the rooks should not have preferred either one. To test the rooks' causal reasoning, the whole tube was manipulated to affect one trap, changing it from non-functional to functional. For Tube C, rubber bungs with holes were inserted into each end of the tube so that the wooden stick could pass through. The only way to retrieve the food was to pull it into the trap without a base, as pulling toward the trap with the raised base would cause the food to become trapped. For Tube D, the whole tube was lowered so that the bottom of the trap rested on a wooden platform. Now, pulling the food toward the trap with no base would lead to the food becoming trapped, while the only successful method was to pull the food across the raised base and retrieve the food at the end of the tube. Out of the seven rooks tested, only one female, Guillem, successfully solved both tubes; all other birds performed randomly. This suggests that causal reasoning is within the capacity of rooks but is still a very difficult task that can only be accomplished by selected individuals.

When similar tasks, but with a different arrangement of colored bases and other color cues, were given to New Caledonian crows, some birds were successful when color was the main cue to correct choice, but not when the base of one of the tubes was now open (equivalent to Tube B). Crows were

Two-Trap Tubes

An adapted version of the classic trap-tube task called the two-trap tube task, can be used to test causal reasoning in tool-using and non-tool-using animals alike. Each version controls for different cues that the subject may use to solve the task. If all these cues are controlled, then causal reasoning becomes the most parsimonious explanation.

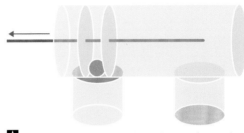

A Tube A requires the bird to pull toward the non-functional trap with the solid raised base, so the food passes straight over and out the tube.

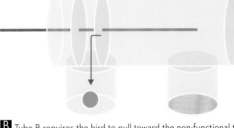

B Tube B requires the bird to pull toward the non-functional trap without a base, so the food falls through the hole.

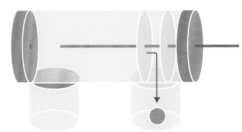

C Tube C requires the bird to pull toward the non-functional trap without a base, so the food falls through the hole.

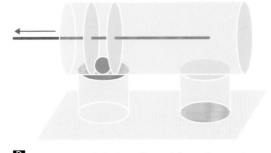

D Tube D requires the bird to pull toward the non-functional trap with the solid raised base, so the food passes straight over and out.

Above A rook correctly performs the Two Trap Tube Task, in which a stick has been pre-inserted into a tube with a functional trap and a non-functional trap at its base (see right). By pulling the stick, the rook was able to pull the food out of the tube without it falling into the trap.

able to transfer their understanding to a conceptually similar trap-table task, suggesting that the crows knew how the task worked, even when it looked very different. However, this seems unlikely as they failed the equivalent task to Tube B. The trap table is conceptually similar to the trap-tube task and so perhaps easier to solve than the two-trap tube task (because there is no distractor trap).

Other birds have been given the two-trap tube task and failed. Woodpecker finches and small tree finches did not solve the initial versions with a pre-inserted tool and only one woodpecker finch solved a tube when it could insert its own tool, but still failed the transfer tube. Keas, cockatoos, and macaws failed the two-trap-tube problems unless they were able to manipulate the food themselves (without having to use a tool), which they did by cheating, pulling the food over the trap with their beaks! So the evidence for causal reasoning in birds remains unclear, since it is based on the results from a single rook and those from a few New Caledonian crows.

Thinking ahead

Human tool use has been suggested as unique because of our ability to use one tool on or with another (associative tool use), or to gain access to another tool that may be out of reach (sequential tool use), or to make another tool more efficient (metatool use). Only sequential tool use has been observed in birds. Among animals generally, the only example of metatool use that comes to mind is the use by some chimpanzees of a stabilizing wooden anvil to make the task of cracking nuts easier.

ONE STEP AHEAD

New Caledonian crows were given the problem of reaching an entrapped food treat when the only tool that could achieve this is out of reach. They were provided with a stone and a short stick that was long enough to reach a second, longer stick but not the food. The crows quickly learned to use the short stick to reach the long stick (ignoring the stone) and then the long stick to reach the food. At first sight this seems impressive, but New Caledonian crows don't use stones as tools, so these objects would never have appeared on their tool radar. Only the long stick was going to be sufficient. If another stick, shorter than needed to reach the food, had been provided as an alternative to the long stick, this would have represented a more rigorous test, one indeed that has since been successfully carried out.

TWO STEPS AHEAD

Captive rooks rapidly learned to use stones and sticks as tools to retrieve food from an artificial apparatus. Rooks were provided with a puzzle box with a tube on top and a collapsible platform held by a magnet, with food placed on the platform. The food could only be released if an object of sufficient weight was dropped into the tube. After a period of rapid learning, four rooks learned how to release the food by dropping various objects into the tube. They were presented with numerous problems in which the type and shape of the objects and the diameter of the tubes were manipulated. They spontaneously inserted stones of different weights, sizes, and shapes, as well as novel sticks of different lengths and thicknesses, into tubes of different widths.

In a test of sequential tool use, rooks were presented with three boxes. Two side boxes had wide tubes: One box contained a small stone capable of fitting a narrow tube, while the other box held a large stone. The central box had a narrow tube and contained a food treat. A single large stone was placed in front of the three tubes. To get the food, the birds had to insert the large stone into the wide tube containing the small stone, then use that stone to collapse the platform of the narrow tube and so release the treat. Most birds performed correctly on the first trial, suggesting that rooks are capable of planning two steps ahead to achieve a goal.

MANY STEPS AHEAD

New Caledonian crows, however, have completed even more complex sequential tool tasks involving three different tools used in order: using Tool 1 to retrieve Tool 2, then Tool 2 to retrieve Tool 3, then Tool 3 to retrieve the reward. Birds were successful in using the tools in series in order to reach the food, but may have used a simpler mechanism than planning in order to do so. In the BBC TV program *Inside the Animal Mind*, a New Caledonian crow was given a multistep problem obliging it to use a number of different tools (see diagram). When observed in a single shot, the crow's behavior looks smart, especially when we learn that this is the first time the crow has experienced all these objects together at once. However, there is more to this feat than meets the eye. The crow had previous experience using all these tools and although it had to put them together in order to achieve its aim, it might have done this via a process called chaining—putting a series of previously learned actions together in an appropriate sequence to achieve a goal. On the surface, the crow's behavior may look insightful, but there is a simpler explanation.

Sequential Tool Use

A New Caledonian crow called 007 performs an eight-step sequential tool-use challenge. However, as smart as its actions may appear, there is a simpler explanation for how he did it.

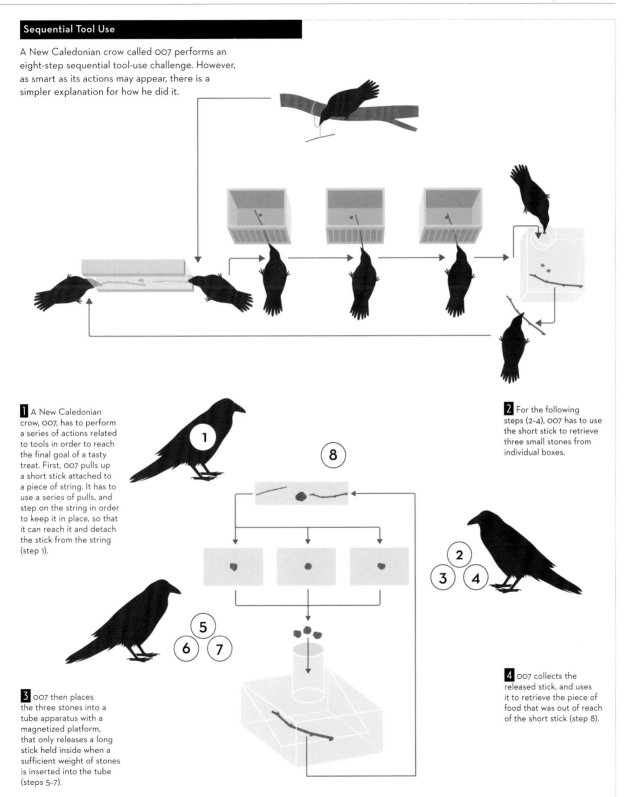

1 A New Caledonian crow, 007, has to perform a series of actions related to tools in order to reach the final goal of a tasty treat. First, 007 pulls up a short stick attached to a piece of string. It has to use a series of pulls, and step on the string in order to keep it in place, so that it can reach it and detach the stick from the string (step 1).

2 For the following steps (2–4), 007 has to use the short stick to retrieve three small stones from individual boxes.

3 007 then places the three stones into a tube apparatus with a magnetized platform, that only releases a long stick held inside when a sufficient weight of stones is inserted into the tube (steps 5–7).

4 007 collects the released stick, and uses it to retrieve the piece of food that was out of reach of the short stick (step 8).

A flash of insight

Insight is the sudden solution of a problem without trial and error. Although animals perform behaviors that look insightful, there are often simpler explanations. For example, chimpanzees provided with an out-of-reach banana eventually stacked boxes underneath to reach it. So rather than using insight, the chimps solved the problem by remembering their previous experience with boxes, combining a series of actions they had already learned. But is there any stronger evidence for insight in animals?

A TOOL-USING NEW CALEDONIAN CROW MAKES A HOOK TOOL...

Researchers in Oxford devised an ingenious task to examine tool understanding in New Caledonian crows. The task was to retrieve a small bucket containing a tasty piece of lamb's heart from a vertical plastic tube. Two crows, Abel and Betty, were provided with two tools made from wire, one a hook, the other straight. Only pulling the bucket up by its handle could retrieve the food. On the first trial, Abel stole the hook, leaving Betty with the straight wire. First, she tried to spear the heart with the wire with no success. After a pause, she inserted the wire into a corner, using it to force a

bend in the wire, thus creating a hook. She inserted the hooked end into the tube and maneuvered the tool under the handle of the bucket, pulling it up and retrieving the food in the bucket.

Was this insight? It is a good candidate despite some reservations. Betty was initially stuck for a solution to the problem when the hooked tool was unavailable, then suddenly appeared to produce a novel method. However, Betty was not born in the lab and so her previous experience of wire and hooks was unknown. New Caledonian crows routinely make hook tools in the wild, so their brains are primed for making hooks. Betty inserted the straight wire into the tube

before attempting to bend it. Not all her hooks were actual hooks, and the technique used to retrieve the bucket possibly could have worked with straight wire. Finally, she was presented with a template hook tool against which to compare.

. . . BUT THEN SO DO NON-TOOL-USING ROOKS

The same experiment was performed on rooks in Cambridge. Wild rooks do not use tools, but have been shown to use stones and sticks as tools in captivity. Four rooks with known history were given a similar tube to Betty, adapted for their larger size and longer bill. First, they were given two wooden hooks, each with a V-shaped crosspiece taped at one end: one with the V facing upward (functional), the other inverted (non-functional). The rooks learned very quickly to use the functional hook to retrieve the bucket. When given a straight piece of wire, three rooks spontaneously bent the wire into a hook and used it to pull up the bucket. Could this be described as insight? Rooks do not make tools; these birds' history was well documented and although they had successfully used wooden hook tools before, those tools were only functionally, not physically, similar to the ones they created. The rooks may have formed a mental image of the task requirements and what makes a functional tool, which they then transferred to the new material (wire) when confronted with this new task. This could be considered insight, though again there are alternative explanations.

Left Rooks in the wild have no need for the use of tools, however in captivity they have shown to successfully perform a number of different tasks that require tools to complete them.

Rook Wire Bending

1 A rook inserts a straight piece of wire into an upright plastic tube and uses the edge of the tube as leverage to bend the wire, thus creating a functional hook tool at the end.

2 As the functional hook is located at the wrong end of the wire to be used as a tool, the rook removes the wire from the tube, flips it over and reinserts the hooked end back into the tube.

3 The rook manipulates the hook to position it under the handle of the bucket holding the food. Once the hook is in the correct position, the rook pulls up the tool with the bucket attached, so gaining its tasty reward.

Fable becomes experiment

One of the famous fables attributed to Aesop is "The Crow and the Pitcher." A thirsty crow comes across a pitcher of water, but the water is so low that the crow cannot reach it with her beak no matter how hard she stretches. But just before losing all hope that she is ever going to quench her thirst, she comes up with the idea of putting stones into the pitcher, thus bringing the water within reach. This story inspired an experiment to test what rooks know about the consequences of their actions.

Rooks were presented with a clear upright Perspex tube half-filled with water, with a worm treat floating on the surface. A pile of rocks of various sizes that could fit into the tube was also provided. If the birds understood that inserting larger stones into the tube displaced more water than smaller stones, they would be more likely to add larger stones to the tube and reach the treat sooner than if they used smaller stones.

WATER AS A TOOL
Four rooks succeeded in inserting stones into the water and three birds quickly learned to place larger

stones into the tube. We wanted to discover whether rooks also knew something about the properties of different substrates compared to water. Water can be manipulated by the addition of heavier and denser objects, such as solids. A solid substrate that acts like a liquid, such as sand (which can be poured), cannot change its condition in the same way as water when a solid is added, because it is a solid itself and so more dense than water. Therefore, adding stones to sand will not raise the level of the sand but will simply rest on the top and so fail to bring the worm treat any nearer. The rooks did place some stones into a sand-filled tube when this was presented alongside a water-filled tube, but they quickly learned to avoid the sand-filled tube and reverted to only adding stones to the water. Rooks are caching birds and so may have been attracted to the sand because of the opportunity they saw to hide the stones in the sand.

The Aesop's Fable task has proven useful for testing whether tool-using and non-tool-using birds understand the causal properties of objects, as well as comparing their understanding with that of human children. A series of clever additions and control procedures have been used with Eurasian jays and New Caledonian crows to determine what they may know about the types of objects that can displace water, such as sinkable versus floatable objects or solid versus hollow objects. In every case, the jays and crows chose the most appropriate objects or substrates, such as tubes with water (not air or sand), larger not smaller stones, sinkable not floatable objects, and solid not hollow objects.

Left In "The Crow and the Pitcher" fable by Aesop, a thirsty crow drops pebbles into a pitcher to raise the water level and allowing it to drink, and is the inspiration behind the Aesop's Fable task.

It has been suggested that the birds that succeeded in these tasks did so not through causal reasoning but rather by adjusting their behavior and tracking the consequences of their actions through the movement of the water. Various tasks were given to crows and jays to investigate this claim. For example, two tubes covered in gray tape were presented, so that the birds couldn't see the rewards or the water in the tube. The only clue to which tube was baited with food was the presence of a large stone in front of the correct tube. The corvids didn't prefer either tube.

DO BIRDS BELIEVE IN MAGIC?

Finally, to determine whether corvids could infer the causal structure underlying a task when not all possible information was available, birds were presented with the U-tube. The U-tube contained three tubes: two wide tubes on either side of a narrow tube placed into a box, so that the bases of the tubes were hidden. Unbeknown to the birds, one of the wide tubes was connected to the narrow tube, which was baited with food. The wide tubes weren't baited, so only placing stones into the wide tube connected to the narrow tube (which was too small to accept any stones) resulted in the water level rising in both tubes and so moving the treat within reach. As most of the tubes were concealed by the box, the birds didn't know two tubes were connected until the water had risen sufficiently after the insertion of a number of stones into the correct wide tube had made the water rise in the narrow tube. None of the corvids chose the correct tube, suggesting they had to see the result of their actions on the tube in order to determine their consequences of what they did.

The Aesop's Fable task has proven one of the best tests for comparing different species in their appreciation of the physical world and the consequences of their actions. Increasingly clever variations of the original task have been employed to tease apart what the birds may have learned and what is the result of reasoning.

Rook Performing Aesop's Fable Task

1 A rook is presented with a tube half full with water, with a tasty worm floating on the surface. But how to reach the worm?

2 The rook starts placing stones into the tube, which raises the water level and subsequently, the worm starts moving into reach.

3 The rook places stones into the tube until the worm moves into beak reach. It takes the worm, and eats its reward. It stops placing stones when a sufficient level has been reached.

IS YOUR CHILD AS CLEVER AS A CROW?

Recent studies have directly compared the performance of crows and jays in various tool-using and tool-making tasks with the performance of young children. These studies were designed to see whether some forms of physical cognition are achieved without verbal reasoning. They weren't devised to say that crows have the same minds as children, as more than 300 million years of evolution separate crows and humans. Indeed, the differences between corvids and children are as interesting as the similarities, especially if one group is particularly gifted or deficient compared with the other.

CHILDREN HAVE PROBLEMS WITH TOOL INNOVATION

The first task used was the wire-bending task. Children of three to nine years old were given a similar version of this task to that given to rooks and New Caledonian crows, but with a sticker as a reward. By age four, children chose a hooked pipe cleaner over a straight one and used it to retrieve the sticker. When children were provided with a straight pipe cleaner, a piece of string, and two short sticks, only a few three- to five-year-olds made a hooked tool and some eight- to nine-year-olds created a novel tool using the pipe cleaner and a short stick. It wasn't until they were much older (sixteen to seventeen years in age) that children could create a hook tool. If four- to seven-year-old children were shown hook making and then provided with the pipe cleaner, sticks, and string, they created their own hook tool from the pipe cleaner. Overall, the lack of insight in children even as old as fifteen was surprising, because children as young as three years old should have had some experience of bendable materials or comparable objects (or the potential for creating such objects).

Other experiments have found that if children are given a pipe cleaner to make into a hook, they are only successful if given a demonstration. If they are then given a hook-unbending task in which they have to straighten a bent pipe cleaner so it fits into a tube and can be used to push out a sticker, they do not use information they learned in the previous experiment to solve the task. Children do better if they are given the direct instruction that they may use the pipe cleaner or string to make something.

Children are capable of remembering the bending task three months later and can use these memories to solve new tasks with novel colors and configurations of the pipe cleaners and buckets in the tubes. If the children are given a conceptually similar task—a piece of dowel with three holes drilled into it and three short pieces of dowel that can fit into these holes—they are not able to create a hook by inserting a dowel into one end, and so transfer what they have learned about hooks from manipulating the hooked pipe cleaner. This is in contrast to the rooks' behavior, which showed that not only were they able to use a wooden hook but that they could then use this knowledge to create a hook from a novel material (wire).

From these studies, we may suppose that young children, until teenage years, are poor at creating new tools, which may be a result of dependence on learning socially from adults. Once the youngster becomes independent, they switch gears and begin innovating their own solutions without support of their close family.

MAGICAL THINKING

The Aesop's Fable task has been used to compare corvids and human children. Like the corvids, children were able to collapse a platform by inserting a weight into a tube on the top of a puzzle box. Once they had learned this, they were presented with a series of tube problems: water versus sawdust, sinking versus floatable objects, and the U-tube apparatus. Across all tasks, performance increased in the older children with peak performance reached in the eight-year olds. From five years old, the children didn't act spontaneously, but like the corvids demonstrated very rapid

Above My niece, Imogen, performing the Aesop's Fable task, putting the correct number of stones into a water-filled tube, to raise the floating toy. Children of varying ages have been tested on a number of variants of this, but it seems only children over eight years old succeed on all the tasks.

Right Despite over 300 million years of evolution since their last common ancestor, there are striking similarities in how crows and humans solve problems. This may be because they had to solve similar problems during their evolution, and these became part of their cognitive phenotype.

learning within five trials. Only eight-year-olds acted spontaneously. Children therefore learned to deposit weights into water-filled rather than sawdust-filled tubes and placed sinkable rubber into tubes rather than floatable polystyrene.

In the U-tube task, which crows and jays did not pass, eight-year-old children learned in a single trial which of the tubes to place their marbles into, while seven-year-olds learned within five trials. All younger children failed. When asked how they succeeded, the children did not infer that there was a hidden mechanism connecting two of the tubes, rather that it worked by magic! They could see that the addition of a marble into the tube had an effect on the narrow tube, but didn't seem to know what was causing it. This magical thinking is common in young children, who start to question how the world works only when they reach a certain age.

It is clear that the capacities of human children have been exaggerated, because rooks, jays, and New Caledonian crows are at least as competent and, in some cases, more competent than children younger than eight years old in tasks requiring tool innovation and causal reasoning. Perhaps our own understanding of the physical world is more closely tied to our use of tools and more reliant on our social nature than that of our feathered cousins.

6 KNOW THYSELF AND OTHERS

Do animals have a sense of self?

When you look in the mirror, you see yourself reflected, but how do you know it's you? You see another human being staring back at you, but why does your brain tell you that the reflection is you and not another human of a similar age, sex, and appearance? Our sense of self is what makes us who we are, individuals in a sea of variety. Our unique physical features differentiate us from every human on Earth, yet our sense of self depends on more than simply recognizing our reflection in a mirror. It arises from our knowledge, our memories, our relationships with others, and our personality, collectively forming the unique barcode that represents an individual life.

MATCHING MOVEMENTS

When asked to comment on our reflection, we don't hesitate to say it's us, unless we suffer from certain disorders of self, such as schizophrenia. But what guides this instantaneous recognition? Are we automatically responding after years of gazing at our reflection, observing changes over time as we grow older, or do we track the movements we observe in the mirror? Can we perceive through our personal body sense that we are the actor whose movements cause the changes displayed on the stage offered by the mirror? If so, is it a simple matter for non-human animals to make the same judgments if they have a similar body sense?

MIRRORS AND MINDS

How do we find out whether animals have a sense of self? Perhaps not surprisingly, mirrors have been used to discover this in the same way as humans. Most animals when first confronted with their reflection treat it as another animal, usually a rival, threatening it by staring back at it. Some animals never learn that the image in the mirror is theirs, always treating it as an unfamiliar conspecific. However, some animals start to treat their reflection differently over time, realizing the movements of the reflected creature mirror their own. If they raise a wing, the bird in the mirror raises a wing. If they open their mouth, the animal in the mirror opens its mouth.

After many hours of mirror exposure, a small number of species start to change their behavior, from socially oriented responses toward their reflection

to self-directed responses using the mirror as a tool to investigate their own bodies. They may look inside their mouths or check their backs or they may perform odd behaviors to see how they appear in the mirror. Gordon Gallup used these changes from social- to self-investigation as the basis for his mark test experiments, initially on chimpanzees. An animal is exposed to a mirror for the first time, with the point at which it changes from social- to self-directed responses, if at all, recorded. Once the animal investigates its own body more frequently in the presence of a mirror than without, the mirror is covered. The animal is then either anesthetized, and given a mark on a part of its body it cannot see without a mirror, such as its face, or it is given a physical mark at the same time as a sham mark that feels the same but doesn't leave a mark on the body, this sham mark being located on a comparable part of the body. After the animal recovers, it is then re-exposed to the mirror, and the frequency with which it touches the physical mark in the presence of the mirror is compared to the frequency that occurs without the mirror, or touching the sham mark in front of the mirror. The animal is said to have demonstrated mirror self-recognition if it either touches the physical mark more frequently than the sham mark or touches the mark in the presence of the mirror more often than without the mirror.

Only a few species respond more to the mark in the presence of a mirror; chimpanzees, orangutans, dolphins, elephants and magpies. As these species are also members of the Clever Club, is mirror self-recognition therefore evidence of intelligence? Is a relatively large brain required for self-awareness as well as intelligence? The jury is still out.

Left A great tit looks at its reflection in a pool of water. Does it recognize that the image looking back is them, or do they just see another bird?

Magpies, mirrors, and marks

The mark test remains the classic way of determining whether an animal has a sense of self. One unresolved issue is whether simpler processes can explain the behavior of the animal in front of the mirror. Does the animal match its self-initiated body movements with those reflected in the mirror, then respond to a novel change—the mark—in the mirror? There is little argument that an animal that touches a mark on its body only when viewed in a mirror notices something different about its reflection. The question remains whether this means it has a sense of self rather than a simpler sense of its own body and position in space.

What do birds do in front of mirrors? Mirrors have been used at Colchester Zoo to elicit mating in flamingos, as they will only breed if their flock is large enough. A parakeet or a finch will spend more time in front of its reflection than in front of another bird or a blank wall. Various crows presented with mirrors have behaved as if in the presence of a conspecific, with females preening and males aggressively displaying in front of the mirror. New Caledonian crows could use the mirror to forage for food that wasn't visible without the mirror, suggesting they understood how the mirror functioned. African greys have also used mirrors to aid their foraging for out-of-sight food. Only three species of birds have been given the mirror mark test, with the only success for magpies.

Different-colored stickers, either yellow or red or black, were attached to the magpies' black feathers. Only the colored marks were visible, but all should have been felt. When initially presented with a mirror before any marks were applied, the magpies produced aggressive displays as if the reflection was another magpie. However, the same birds were seen to prefer a chamber with a mirror over one without. Next the magpies were given a mark test. On first exposure to the mirror after being given a visible mark, five birds spontaneously displayed mark-directed behavior that was more frequent than when there was no mirror or mark. However, the magpies did not try to remove the black mark, suggesting they discriminated between marks based on what they could see in the mirror. But they did successfully remove the colored marks and then immediately ceased their self-directed behavior. Finally, not all magpies behaved this way and indeed some birds remained socially responsive to their reflections rather than seeing these images as reflections of themselves.

A BRIEF NOTE ABOUT PIGEONS AND MIRRORS

One argument against the mark test is that self-directed behaviors very similar to those displayed by chimpanzees can be achieved through careful training of individual components of the behavior and are therefore nothing to do with self-recognition. Robert Epstein laboriously trained pigeons to peck at dots originally projected onto walls then presented on their chests (hidden through wearing a neck brace, so only seen in a mirror) until their behavior resembled the self-directed behaviors of chimpanzees. However, this study hasn't been replicated and so probably says little about what may be going on when an animal naturally views its reflection without rewards or training.

Left A magpie with a yellow sticker placed on its throat. The sticker contrasts with the black feathers, and should easily be seen reflected in a mirror. A self-aware magpie should notice this change from its normal appearance, and trying to displace the sticker.

AN ALTERNATIVE TO THE MARK TEST

The mirror mark test is not without its critics, but there are few alternatives. Testing an animal's appreciation of self is extremely difficult, because we cannot ask them what they think. An alternative task also uses mirrors, not in the concept of self-exploration but with regard to differentiating self and other. Many caching birds protect their food prize from thieves. One method is return to their stashes when others have left the scene and move them to new places the original viewer has not seen. Western scrub jays hid food either when alone or in the presence of a prospective thief or in front of a mirror. When the jays returned to their caches at a later time alone, we predicted that if they thought that the mirror was their reflection, they should have treated their caches in the same way as if they had been alone (that is, not under threat). However, if they thought that the mirror was another jay, they should have treated their caches as if they had been watched by another bird (that is, moved them to new places unknown to the spy). They did the former and left their caches alone. This suggests that the scrub jays may have realized either that it was their reflection looking back at them and so they did not need to protect their caches or that the scrub jay in the mirror behaved so strangely that they did not need to act protectively. Further experiments are needed to tease apart these ideas and test whether other corvids have some (rudimentary) appreciation of self, perhaps using new methods.

Mirror Self-Recognition by Magpies

A magpie without a mark inspects itself after being exposed to a mirror (1). The same magpie is exposed again to the mirror after receiving a yellow mark on its throat (2). The magpie tries to remove the mark by self-preening with its bill (3), foot (4), or neck/wing (5).

1 Exposure to a mirror

2 Exposed to mirror with mark

3 Self-preening with bill

4 Self-preening with foot

5 Self-preening with neck/wing

MENTALLY TRAVELING IN TIME

We spend much of our lives contemplating our past and thinking about our future. We have already encountered the idea of mentally traveling into our specific past through episodic memories. We can also mentally travel into possible futures from planning what we're going to have for dinner, to our summer holidays, to longer-term plans for our retirement and even our funeral.

IN OUR MIND'S EYE

This planning requires an imagination as the future hasn't happened yet and could take one of many alternate paths. We have to think in our mind's eye what a probable future will be like and plan accordingly. We might, for example, want to eat a certain food for breakfast, such as cereal, but want something different for dinner, such as steak. If all we bought when we visited the grocery store was breakfast cereal, we would get bored very quickly and wouldn't fulfill all our nutritional needs. We may also shop for groceries when we are sated, having just eaten, but we still have to plan for a time in the future when we will be hungry again. Does this ability require language, is it shared with non-human animals such as birds, and is it dependent on culture? Were we capable of planning for a long-term future in the time before shopping lists and refrigeration?

One problem with attempting to find out answers to these questions is that animals don't use human language, and most of our tests for humans are based on language. We therefore have to think about which behaviors are future-oriented and which require thinking about the future. Some animals behave as though they appreciate future needs, by, for example, building a nest or home for a harsh winter, hiding food for future consumption, or migrating to a warmer climate, but there is little evidence that future thinking is required for this behavior. These appear to be innate behaviors displayed by all members of a species and triggered by changes in the environment, such as reduction in temperature, or hormonal changes, such as occur during a breeding season.

FUTURE MEMORIES

It is impossible to test for conscious awareness of a future time in non-human animals. Humans appreciate that we are the authors of our plans, that we plan for our specific future not in general response to current conditions.

This form of conscious awareness is impossible to test in non-linguistic creatures, so experiments are designed that manipulate strict behavioral criteria, the outcomes of which can be seen and changed based on experimental conditions. As with experimental demonstrations of episodic memory, there are three behavioral criteria for demonstrating future planning in animals: content, structure, and flexibility. Content refers to the individual components of a future event: what will happen, where, and when in the future. These contents are bound together in a cohesive structure that is similar to an episodic memory (but in the future). Plans for the future have to be flexible depending on differing circumstances that require updating. For example, your refrigerator breaks down, so you need either to eat the food sooner than you had planned or to move it somewhere else that's sufficiently cold.

IF I'M HUNGRY NOW, WILL I BE HUNGRY LATER?

One arguable criterion for future planning is that behaviors have to satisfy the Bischof-Köhler hypothesis, namely that any future-oriented behavior has to be for a future motivational state not a current one. If an animal is hungry now and acts to satiate that hunger, then its behavior cannot be oriented to a future state of hunger. This criterion is slightly problematic. Future-oriented behaviors are not just about motivational states. Everything I do that is geared toward the future isn't just to satisfy my hunger, thirst, or sexual appetite and it's not clear why animals have to be chained to their motivations, either. It is also possible to possess two competing motivational states running concurrently or even different levels of the same motivational state. This is clear from the phenomenon of specific satiety. I can be generally hungry, but when I have eaten my fill of steak, that doesn't mean that I am satiated on all foods, only steak. I may remain hungry for a different food, such as ice cream.

The Bischof-Köhler hypothesis ignores these possibilities. However, recent experiments with scrub jays have found that they can plan to cache and eat one food when they have been satiated on another, thereby caching for a future need. The scrub jays may be satiated on food A, but will cache food A in locations in which they will be able to retrieve it later when they have become satiated on a different food B, but are hungry once more for food A. Jays can therefore prepare now, in a state of satiation, for a different motivational state in the future, namely, hunger.

Above and right An Eurasian jay stuffs acorns into its throat pouch—which can hold up to seven acorns—each one will be hidden in different locations, usually next to tall landmarks, such as trees (or even in trees [right]), in order to provide a store of food that will be available months into the future when food is scarce.

Planning for breakfast

Western scrub jays hide food for future consumption, but remember specific past events related to caches they made in the past, so-called what-where-when memories. In human psychology, episodic memory has been tied to future planning where our past directs our future. Caches are made in the present to sustain a need in the future. This future need can be a motivational state, such as hunger, or an environmental need, such as food becoming scarce in the future (for example, the following winter).

A HUNGRY FUTURE

Scrub jays cease caching in a specific location if those caches are stolen or become degraded in the future. A study was designed to examine whether scrub jay caching was oriented to a future motivational state. Scrub jays were temporarily housed in a set of three interlocking compartments (A, B, C). A caching tray was placed in each of the two outer compartments (A, C), while the central compartment B contained a bowl of powdered pine nuts (which cannot be cached). Over the course of the next six days, the jays were trained so as to appreciate the consequences of being placed in compartments A or C the next morning.

Every morning, the jays were randomly placed in either compartment A or C (three times in each). In compartment A they were provided with a bowl of powdered pine nuts, whereas in compartment C they received no breakfast (but received brunch later). During the next six days, jays were either allocated to the Breakfast Room (3 days in compartment A) or the No Breakfast Room (3 days in compartment C). They had no idea which compartment they were going to be

placed in the following morning, so didn't know if they were going to be fed breakfast. On the evening of the seventh day, rather than being given a bowl of powdered pine nuts, they were given a bowl of whole pine nuts that could be cached and provided access to all three compartments. If the jays understood that they could be placed into the No Breakfast Room the following morning, they should plan for this eventuality by caching food in the tray in the No Breakfast Room; they wouldn't need to cache in the Breakfast Room because breakfast would be available the next morning in that room. This is more or less exactly what the jays did, caching mainly in the No Breakfast Room, suggesting that they knew to pre-empt the lack of breakfast the next morning.

WHAT DO I WANT FOR BREAKFAST?

In a similar experiment, rather than allow the jays to eat breakfast in one compartment or go hungry in the alternative compartment, they were provided with kibble in one and peanuts in the other. Over six days, the jays learned which food would be served in which compartment the following morning. On the evening of the seventh day, the jays were provided with both kibble and peanuts (whole) that could be cached in either compartment. The jays cached peanuts in the Kibble Room and kibble in the Peanut Room, as they preferred a choice of foods for breakfast. This behavior supports the future planning account; they were thinking about what would occur at breakfast the next morning. Their behavior in the evening was dependent on their future motivational state the next morning.

Left It is expected that because scrub-jays hide perishable foods that decay at different rates, and they cache in front of potential thieves, they understand that different conditions at the time of caching may influence whether their caches will be available when they come to recover them.

Planning for Breakfast

In this planning for breakfast experiment with Western scrub jays, a scrub jay can search three interconnected compartments (A, B, and C) and eat powdered pine nuts. Compartments A and C contain caching trays. Over the next six mornings at 7 a.m., the jays are randomly confined in either compartment A (the Breakfast Room) or compartment C (the No Breakfast Room), where instead they will get brunch at 11 a.m. On the evening of day 7, they are given whole pine nuts that can be cached in either compartment A or C. If they can think about what might happen the next morning, they should cache pine nuts in compartment C, as they won't be served breakfast in that room.

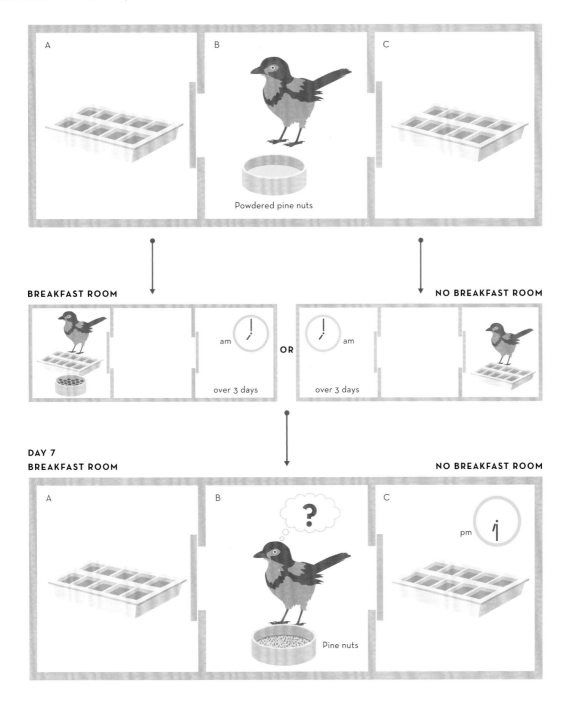

Powdered pine nuts

BREAKFAST ROOM

NO BREAKFAST ROOM

am
over 3 days

OR

am
over 3 days

DAY 7
BREAKFAST ROOM

NO BREAKFAST ROOM

Pine nuts

Thinking about other minds

The idea that some animals are self-aware is controversial, but the idea that some animals understand what others think and feel has been argued over just as strongly.

THEORY OF MIND

Humans find it easy to read minds and do so from an early age, not in the sense of a TV-style "mentalist" but by picking up subtle cues as to the state of another's mind. You know someone is lying by the way they avert their eyes when speaking and you know when someone is attracted to another by the duration of their gaze. Young children know when someone is pointing something out versus pointing because they want something. This ability to comprehend the content of another's mind—what they perceive, what they know, what they want, and what they believe—has been termed theory of mind and is suggested to be uniquely human.

THE ADVANTAGES OF BEING A MIND READER

It has been proposed that only humans form theories about the content of other minds, either based on our own understanding of the world or through natural precursors, such as being able to follow where someone is looking or pointing. Despite this, we can only be certain about the content of our own mind,

hence the reason for calling it a theory. Knowing what others can see, may have seen, and may believe they have seen (which may be incongruous with reality) gives a social being a distinct advantage over others that don't possess this skill. Creatures without a theory of mind would need to learn about each individual in their social group through repeated, numerous interactions in order to predict their behavior, whereas an individual with a theory of mind would be able to use a set of rules to do so on the first occasion upon meeting someone new.

MAKING PREDICTIONS

Although the advantages of possessing a theory of mind are obvious, the argument remains as to whether animals need to think about other's minds or simply process information about another's behavior. I can predict what you will do next, with a high degree of accuracy, based on following your line of sight, the direction in which you are walking, and where you are reaching. I do not necessarily need to invoke concepts of seeing or intention to make a simple prediction,

Left A pair of ravens at the Tower of London. One raven is caching food, while the other is distracted and cannot see where the cache has been made.

but simply follow the path of your movements until they reach a potential goal. I don't need to know why you have behaved as you did, only use your behavior to make predictions about what you will do next. Reading your behavior may be sufficient to tease apart simple contents of your mental life, but some mental states, such as beliefs, are not so easy to decode.

MIND-READING BIRDS

Studies on animals, including birds, have investigated whether they can use simple behavioral cues, such as another's eye gaze direction, pointing gestures, or what they have or haven't seen in the recent past as precursors for testing whether they have an appreciation of another's visual access (can they see?) or their knowledge (do they know something but are they ignorant of something else?). Despite almost

forty years of study, the idea of theory of mind in non-human animals remains highly contentious—and not without good reason, because it is still unclear how the human theory of mind works and whether it is as sophisticated or even as ubiquitous as previously thought. Some of the studies on birds have provided tantalizing data that puts mind reading within a strong evolutionary framework—thinking about the ecological circumstances in which a theory of mind would be adaptive, such as protecting food caches, and how it may have evolved.

Below Although birds, like these titmice, may not be capable of humanlike theory of mind, they demonstrate sophisticated social skills allowing them to respond appropriately to others' social signals, and even predict their behavior. These skills, however, are likely the product of innate responses, genes or rapid learning, rather than an appreciation of other's mental states.

Another's desires

Most birds form monogamous relationships and some partnerships can last throughout the birds' entire lives. These pair bonds are dependent on the couple working together and potentially second-guessing their partner's desires in order to make an efficient team. This provides an advantage in protecting and raising a brood of healthy offspring.

VALENTINE'S DAY

Food sharing is a behavior at the heart of many avian pair bonds. Some males offer potential partner nuptial gifts to entice them or they may make offers of food to others as a display of their ability to provide or their potential as a cooperator—two essential components to being a good father or partner. Continued gift-giving may help maintain a partnership that isn't set for life. By determining their partner's desires early on, male birds can tailor their gifts to their partner's particular needs. Your partner's motivational state can also have a strong effect on his or her desires. If your partner has just eaten chocolates, he or she may not want any more but may still be happy to receive some fruit. Does the decision of what to offer depend on seeing what your partner has just eaten? Can birds recognize others' desires, not just pick up simple behavioral cues, such as staring longer at one food than another?

JAYS BEARING GIFTS

Eurasian jays form pair bonds throughout the breeding season but don't maintain those bonds year upon year. Male jays share food with their female partner. A male was fed his normal diet until he could eat no more (was satiated) and observed his partner eating her normal diet, either wax worms or mealworms, each on different days. If the female had been sated on her normal diet, then the process of specific satiety suggests that she would be equally interested (if at all) in eating a new food, such as mealworms or wax worms. However, if she had previously eaten wax worms, she should then have preferred mealworms and, equally, she should have preferred wax worms if she had previously eaten mealworms. The male, who was in the next aviary to his partner, was offered a single mealworm or wax worm and could either eat it himself, cache it, or share it with her. If the male shared, he tended to do so based on what he perceived to be the female's desire—the food she hadn't eaten—so gave her a mealworm if she'd eaten a wax worm and vice versa.

Did the female provide a clue in her behavior that led to the male's choice to present her with a specific type of food—for example, begging when she saw the male presented with the food she wanted? Many birds produce a begging call and gesture when they want to be fed. To control for this possibility, the female was fed out of the male's sight so that he would not know what she had been fed and could only choose what to feed her based on her behavior. The males failed to feed the female based on her desire, suggesting that they had based their earlier decision on an understanding of her desire for a specific food that she hadn't eaten rather than a signal for a specific food. This behavior could be based on introspection ("I've eaten X until I was full, then I wanted to eat Y") or desire attribution ("she had been given all of X that she wants, so now she will want Y"). At present, it is difficult to distinguish between these processes, but both suggest the males responded to what they perceived to be the females' state of need.

Left A male Eurasian jay knows what his female partner might want to eat, based on observing her eat a particular type of food until she wanted no more. He feeds her the food she should want, and avoids giving her the food she's already eaten.

Desire Attribution

A desire attribution experiment in Eurasian jays. A male jay either observes a female eating mealworms (M) or wax worms (W) until she is full (1) or his view of her eating is blocked (2). When the male is given a choice of a mealworm or wax worm, which does he feed his female, based on his previous viewing experience?

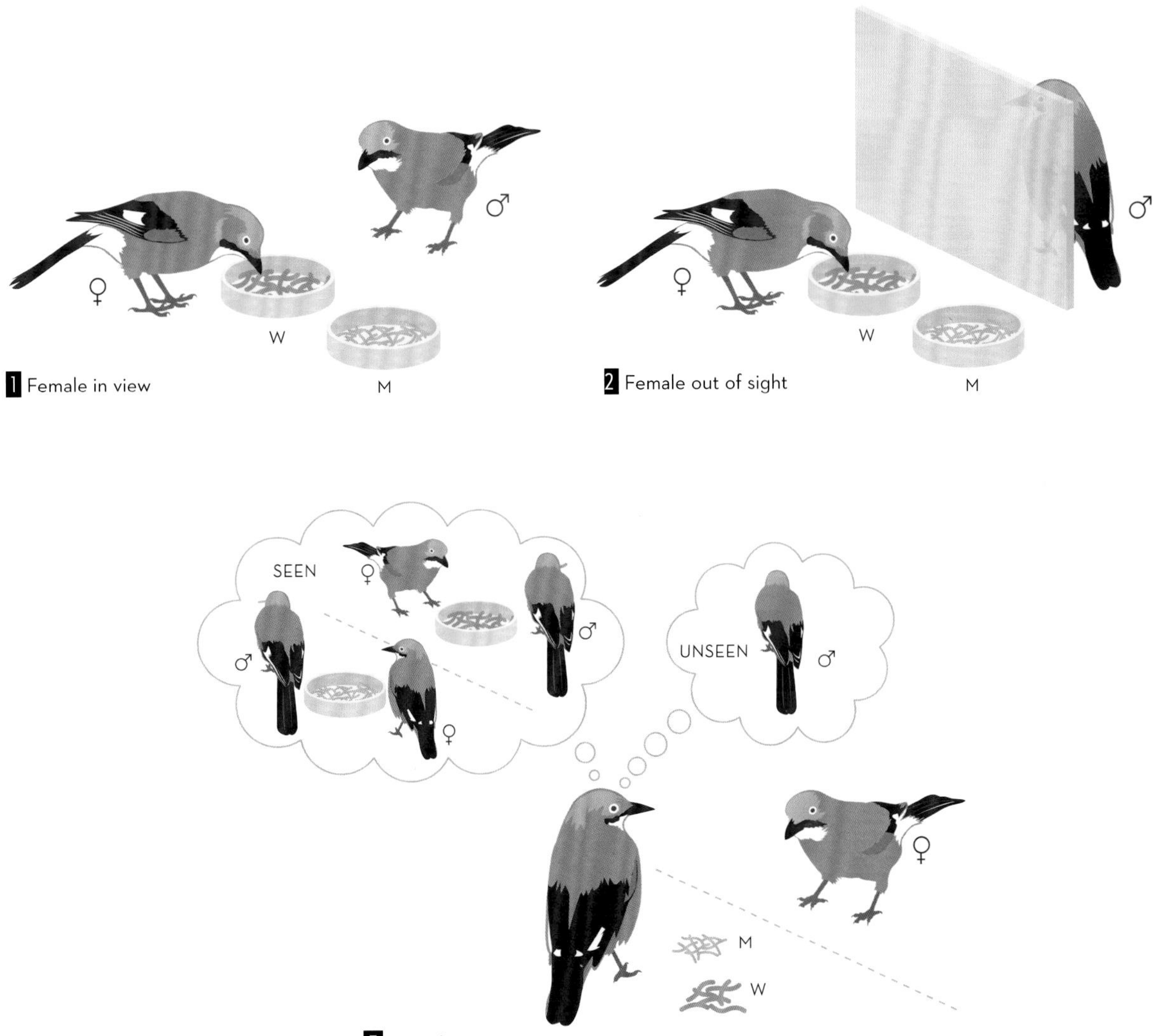

1 Female in view

W

M

2 Female out of sight

W

M

SEEN

♀

♂

♂

♀

UNSEEN

♂

M

W

3 Test phase

CACHE-PROTECTION STRATEGIES

To birds that hide food for future consumption, be it for later the same day because they cannot carry all they find, or for the harsh winter months sometime in the future, food stores can mean the difference between life and death. It is just as important to protect caches from thieves as it is to remember where they were hidden. Food-caching birds implement a number of different strategies to protect their stashes from potential pilferers, some requiring sophisticated socio-cognitive skills.

SOCIAL STORERS

The problem of protecting caches is most acute for social birds that may find it difficult to hide food without others watching. But while waiting for potential thieves to leave the scene may not be viable, there are strategies they can employ, such as moving caches around, to make the final hiding place difficult to determine. This strategy may have occurred accidentally in the past and resulted in a greater yield of recovered caches, or it may be the result of a deliberate attempt to deceive thieves.

Social cachers appear to distinguish between the different identities of those observing them and use different strategies based on this discrimination. For example, if their partner is watching, they don't use protective strategies because they are planning to share caches with their partner. If the observer is of a lower rank, they also don't protect their caches, which is in contrast to their reactions to a higher-ranked dominant. This is a good strategy in some respects because, unlike a dominant, a subordinate will not attempt to steal caches when the storer is present, though they may indeed be more likely to use surreptitious means to steal when the storer has left the scene.

THIEVES AREN'T THICK

Experiments on the cache protection strategies of corvids have found great flexibility in their use of different strategies, dependent on the social context during caching (that is, whether another bird of the same species is present or what is the identity of the observing bird) but also on the extent to which they can see the caching event. The concept of seeing is controversial because it is a mental state, so if an animal can appreciate that another can see something, then they ought to possess a rudimentary theory of another's mind (at least in terms of the mental state "see"). However, this is not clear-cut because although we would attribute seeing to an animal with its eyes open and oriented toward the caching event, there are alternative, simpler explanations to describe the viewing bird's actions that do not have to appeal to understanding its mental state. It may have learned, for instance, that it loses more caches if these are made in front of a conspecific facing it with eyes open when the cache is being made than when that individual either has its back turned or isn't there. This doesn't require a concept of seeing but would lead to the same protective behavior. This response requires the cacher to learn (perhaps over hundreds of trials) the result of caching in the presence of this particular stimulus. It must also generalize across other configurations of head, eyes, and body position in relation to the caching event, which could number thousands of variations. This is not very satisfactory, so an alternative explanation is that the clever animal has formed a rule that can be generalized across many different scenarios, such as "If X is oriented toward the caching event, then caches are more likely to be stolen" and "Caching when X is not oriented toward the caching event will lead to more caches remaining safe." Some psychologists have argued that this is all that is required to explain the bird's behavior, but lab

Right Acorn woodpeckers, as their name suggests, depend on acorns for food. These are cached for the winter in small holes in tree trunks and as the acorns dry out, they are moved to smaller holes in the tree.

studies on hand-raised birds using a small number of novel trials do not provide much opportunity for the birds to learn such complex rules within the short time available. In many cases, either they do not experience their caches being stolen or any stealing is unpredictable, so the opportunity to learn is limited.

Food-caching corvids appreciate the difference between competitors that are present and those hidden behind barriers, between those that are located far from cache sites and those that are located close by, and between those cache sites that give full visual access because they are fully illuminated and those that are hidden in shade. In all three cases, a rule based on the visual access available to the viewer—whether it can or cannot see—is sufficient to implement a protective strategy, whereas developing rules based

on learning would be inflexible and require fresh adaptation to each new situation. With only three trials for the bird to learn each new rule, and often with correct performance achieved on the first trial, this scenario seems implausible. Sometimes explanations based on supposed simpler mechanisms such as trial-and-error learning may be more complex than those based on cognition or mind reading.

Below Caching birds, such as blue jays, are not only protective of their caches from potential thieves of their own species (conspecifics), but also other avian or mammalian thieves (heterospecifics), such as woodpeckers and squirrels.

Cache Protection

Different cache protection strategies employed by food-caching birds to counter theft by potential pilferers.

1. Consumption
2. Enhanced caching
3. Reduction
4. Cessation
5. Delay
6. Spacing
7. Out of view
8. Difficult to see
9. Multiple moves
10. Re-Caching

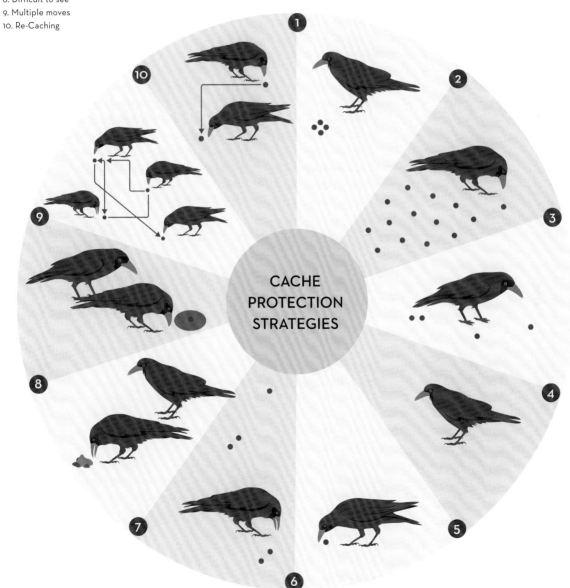

Knowing what others know

If we assume some food-caching birds appreciate that competitors can see certain caching events but not others (because their view is obstructed or degraded because of distance or low light levels), can we then go a stage further?

SEEING LEADS TO KNOWING

To know something, you have to perceive it. If I witness an event, then I will have knowledge of that event. If I wasn't present at the event or my view was blocked, then I will be ignorant of that event. Knowledge in this context is the retention of sensory information that guides subsequent behavior. If a thief witnesses a bird caching in area A, then it should possess knowledge about the cache's location and, if the cache isn't moved, it can use such knowledge to steal food from that same location. If the storer also has knowledge about the thief (that is, that it was present during caching in area A) then the storer may attribute not just perception (awareness of the current state of the world) but also knowledge (memory of the past state of the world) to that other bird even when it is no longer present.

The storer may understand that though the thief may no longer see the caching event, it may still retain knowledge of it. However, it also knows that individuals not present at the caching event remain ignorant of it.

CAUTIOUS RAVENS

Ravens have been tested to determine whether storing birds understand that potential thieves have specific types of knowledge about certain events. At the simplest level, can cachers differentiate between individuals who witnessed caching and individuals who did not? If a raven stored food in the presence

Below Rooks cache food, but because they are highly social, have to do so under the beady eye of their flock mates, some who might steal their caches. To protect their hoards, rooks either cache as far away as possible, bury their caches deep or use their partner as a look out or distraction.

of an individual who witnessed the caching and another whose view was blocked, when all three individuals were released into the caching arena, the storer was more likely to recover its cache if the individual that had witnessed the caching approached the hidden cache. Storers tended to hold off recovering caches in the presence of ignorant birds, so as to not provide them with information they did not already possess.

Pilferers tested with competitors who either witnessed caching or were not present during caching may also attribute different levels of knowledge to these individuals. If a competitor was dominant but had not witnessed caching, the observer delayed pilfering, because presumably, if it stole the caches in front of the ignorant dominant, the dominant would now know where the caches were and so steal them from the weaker bird. If a dominant competitor had witnessed caching, the observer quickly pilfered the caches so as to be the first to gain the prize. If a subordinate competitor had witnessed caching, the pilferer also rapidly retrieved the caches, because there was no reason to delay, there being no competition for them once they had been recovered.

This suggests that both caching and pilfering ravens appreciate that different individuals who have seen different events use different behavioral strategies to protect or steal those caches as required. One explanation for these differences could be that ravens recognize different individuals possess different states of knowledge. This could be based on their presence versus absence at an event or their behavior after the event, such as approaching or ignoring the cache site.

Raven Knowledge

Upper: A experimenter hid two pieces of food in the presence of a raven (observer in the left cage; O), whereas a second raven (non-observer in the left cage; NO) could not see the caches being made. A third raven (focal subject in the right cage; F) could see both caches being made. Lower: Either O or NO was allowed to view the caches. F was released shortly before O or NO, and had to decide which cache to retrieve based on whether their competitor had seen the caches being made or not. In an alternative condition, O and NO each saw a different cache being made, so F had to direct their efforts onto the specific cache their competitor had seen.

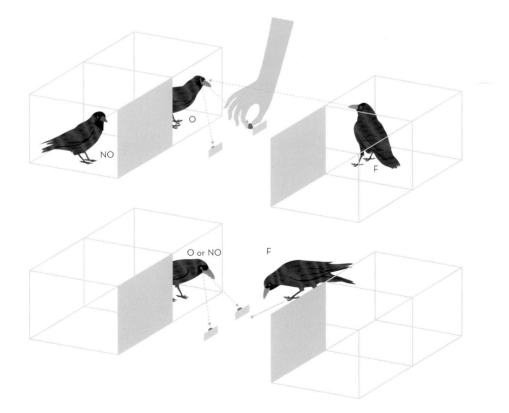

DO BIRDS DECEIVE ONE ANOTHER?

The ability to deceive another, and so exploit their resources, such as take their food or mate with their partner, has been suggested as a primary driver for the evolution of intelligence in primates, as it would seem to be dependent on understanding another's mental states, and manipulating them.

TRICK AND TREAT

Like ravens, scrub jays differentiate between jays with different states of knowledge. A jay was provided with two sand-filled trays in which it could cache wax worms. One tray was blocked with a transparent barrier while the other was open. A competitor, observer A could see caching in tray A, but who was switched to observer B when tray A was blocked, and caching occurred in tray B. When both trays were returned three hours later, the storer was allowed to recover its caches from each (now both open) on one of four conditions. It could recover them either in view of observer A or observer B, or in private (no other bird present), or in the presence of a third bird, observer C, which had not witnessed any caching in the trays. Storers were very selective about where they re-cached food from the two trays, only re-caching items from tray A when in the presence of observer A and only items from tray B in the presence of observer B. They re-cached items frequently from both trays when in private and few, if any, from either tray in the presence of the new observer C.

We might suppose that re-caching from both trays when in private makes sense, as caches would be moved from locations that were known to the previous observers to locations subsequently unknown. The absence of re-caching from the trays in the presence of the new observer C also makes sense, because this bird had not been given earlier access to information about the presence or location of cache sites having not been present at the time of caching.

Below A Western scrub-jay caches in a sand-filled ice cube tray. As the bird is being watched, it removes the cache multiple times, moving it to different locations in the tray. It may even appear to have removed a cache, but left it buried. This deception is similar to the human shell game.

CONFUSING THE COMPETITION

What might seem odd is why storers would re-cache food at all in the presence of observers. Two points are relevant. First, they only re-cached items from the trays that had been seen to be cached, not from the other tray; and, second, they re-cached some items multiple times. Whereas items that had been re-cached in private were moved to new locations and then only once, items that had been re-cached from observed trays were re-cached to old cache sites on many occasions, even up to six times. Moving a cache to a different location multiple times might serve to confuse the viewer, especially if the cache wasn't left in the last place in which the storer seemed to have left it. Video analysis revealed that the storers sometimes left a cache in the second, third, fourth, or even fifth locations, but not the final sixth. This behavior may act like the traditional shell game.

THE SHELL GAME

For hundreds of years, charlatans have used a traditional gambling game-turned-confidence trick to swindle money from unsuspecting punters. The con artist presents the audience with three cups or shells and a ball is placed under one of them. The cups are moved around and the punter has to keep his or her eyes on the cup covering the ball. Inevitably, the punter cannot track the ball to its final resting place because the con artist cheats, placing the ball wherever he or she wants it to be at the end. The scrub jay plays an equivalent game to the shell game when it moves its re-caches multiple times without necessarily placing them in the final location in which they seemed to have ended up. This may serve to confuse the viewer. The fact that the storer occasionally retains the cache in its bill or throat pouch means that the viewer is not party to all the information it needs to successfully retrieve that cache at a later time.

It takes a thief to know a thief

Western scrub jays use different strategies to protect their caches from potential thieves. One predominant strategy, re-caching, involves moving caches from their original place in view of another to a secret location out of their sight. What controls re-caching?

THINKING LIKE THIEVES

Scrub jays cached in a unique sand-filled tray, surrounded by colored blocks, in the presence of another jay and then in another tray when in private. After three hours, they were enabled to recover all their caches in private. They selectively moved caches from the tray in which they were seen caching to locations in a new tray that would not be known by the original observer. They did not remove and replace caches from the tray in which they had previously cached in private.

Re-caching is a strategy used by all scrub jays, but is it learned, innate, or part of a mind-reading system? An experiment aimed to find out by raising young jays in three groups with different experiences. The observer group was given the experience of observing another bird cache, but was never allowed to steal those caches. The observer and pilferer group was also given the experience of observing another jay cache, but could steal those caches. A final pilferer group could steal any caches they encountered in trays, whether they had made them or not. The three groups then cached in distinct trays, one tray in the presence of another jay, another tray when caching alone. Birds in the observer and pilferer group responded by selectively re-caching from the observed tray into new places, but not from the tray in which they had cached in private—which

was not at risk. The pilferer group behaved similarly, so even though they hadn't been given the chance to observe another jay caching (at least in the context of the experiment), they still used a successful cache protection strategy. By contrast, the observer group displayed very little re-caching from either tray.

EXPERIENCE PROJECTION

This difference between groups supposes that the experience of pilfering another's cache (that is, those not made or observed being made by the thief) is sufficient to elicit cache protection behaviors; it takes a thief to know a thief! Birds without pilfering experience do not use anti-theft strategies. This difference between birds with and without experience of pilfering has been proposed as experience projection. When birds steal for the first time, they retain this experience or memory in their brain. It is not clear what form this memory takes, only that it relates specifically to the potential scenario of caches being stolen. This memory has to be specific to theft by another individual (of the same or a different species) and not a general idea that caches can disappear or degrade through natural processes. It is not known whether observing theft is sufficient to form strong enough links in the brain to initiate the anti-theft brain program. When birds in the pilfer group are given the opportunity to protect their own caches, the brain program is triggered and they project their own experience (of pilfering) onto another jay that has the potential to be a thief (that is, the observing bird). They initiate re-caching as a protective strategy. This has been described as a form of introspection or simulation—attempting to model another's actions or intentions based on your own experiences—and may be the best case of theory of mind so far observed in animals.

Left In lab experiments, scrub-jays get to cache either when there are no prying eyes around, or when another scrub-jay can see their caches. In the latter case, they need to implement a number of strategies to reduce the chance that the potential pilferer will successfully steal their caches later, such as hiding them in darkened corners or behind barriers.

Experience Projection

A Western scrub jay caches wax worms in one tray when observed (the observed tray) and a different tray when alone (the in-private tray). After three hours, the bird is given back either the observed tray and a new tray or the in-private tray and a new tray, and allowed to recover its caches. What does it do with its caches? Groups with experience as thieves (observer and pilferer group and pilferer group) move and hide (re-cache) food from the observed tray to a new tray, but only when they have been observed caching. Birds without experience of being thieves (observer group), do not do this.

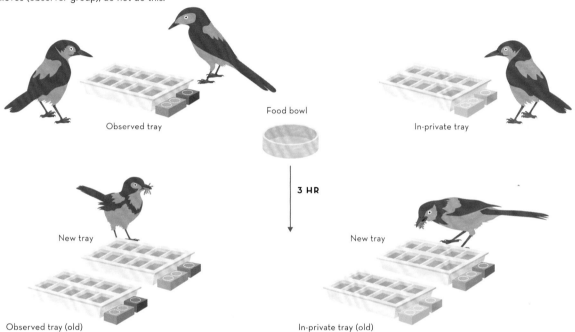

Observed tray

Food bowl

In-private tray

3 HR

New tray

Observed tray (old)

New tray

In-private tray (old)

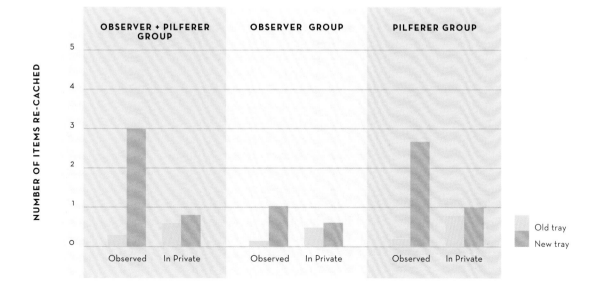

Walking in another's shoes

If you see someone in pain or suffering from loss, you feel sympathy for him or her. However, if you have experienced the same pain, suffering, or loss, then you will feel empathy.

SHARING YOUR PAIN

When feeling empathy, you do more than see the effects of a person's emotional distress,; you have experienced those emotions yourself. You know exactly how the person is feeling, because you have previously walked in his or her shoes. To feel empathy for another living being, you need to look inside yourself and examine your experiences to extract information about this other individual's thoughts and feelings. This is done automatically; you cannot help but experience the individual's pain and distress.

ANIMAL EMPATHY

The study of animal empathy is relatively recent and unfortunately based largely on conjecture and anecdote rather than hard evidence. The evidence does not differentiate between our closest relatives, chimpanzees, and other animals such as mice and birds. Methods to test for empathic responses are either to cause distress in one individual and observe the reactions of those around or to observe naturally occurring events that cause distress, such as one animal being the victim of another's aggression. Do bystanders try and comfort the victim? This affiliative behavior after an aversive event has been termed consolation, although this term is controversial being named after its supposed function rather than what it can be categorized as doing (that is, third-party post-conflict affiliation). Empathy should be more common in individuals that care or protect one another, such as a parent with their offspring or one breeding partner with his or her mate.

COMFORTING CHICKENS

In domestic chickens, mother hens and chicks were exposed to a distressing event—air puffs to the face. The hens were fitted with a heart-rate monitor and either received air puffs themselves, or their chick received air puffs, or there was no treatment. While the hens' heart rate did not change when they received a

Above Despite ravens having strong bonds, fights do occur, especially between pairs over food and power. Bystanders to the fight, especially if their partner was involved, console the victim, suggesting a possible empathic response to another's distress.

series of air puffs themselves, it did increase when their chicks experienced this distressing event. The amount of time they spent standing alert also increased while the time spent preening decreased, suggesting their attention was more focused on their offspring. However, it is hard to say conclusively that the mothers were empathic toward their chicks rather than simply more aroused or alert to what their young were enduring, because their own heart rate did not change when they received air puffs. As the hen did not experience distress due to the air puffs, it is unlikely that she was experiencing the same feelings as her chicks.

A FRIEND IN NEED...

Very few species demonstrate consolatory behavior after a fight. The PC-MC method discussed in relation to reconciliation can also be used to examine the prevalence of consolation. Affiliative behavior between the victim and a bystander is recorded in the ten-minute period after a fight, though not between the victim and aggressor. This is the post-conflict period (PC). All affiliative behaviors that occur with these same individuals the following day, at a similar time are then recorded in the matched control period (MC). A comparison between the PC and MC periods demonstrates whether consolation has occurred. Usually, there is a spike in affiliative behaviors in the first couple of minutes after the fight, but this tails off quickly. After a fight, there tends to be a lot of close physical contact. In chimpanzees, this is expressed as embracing and kissing, whereas in rooks it is bill-twining (avian kissing) and pair displays. In chimps, post-conflict behavior occurs between any individuals, especially those with strong relationships or familial ties, whereas in rooks it is primarily between partners. In ravens, as with rooks, bystanders associate more with the

victim after a fight than during an MC period with no aggression, and those directing their affiliation toward victims are more likely to have a valuable relationship with them. At the extreme end, those individuals will be the victims' partners, as is the case with rooks.

The suggestion of empathy in the consolation behavior of corvids is made stronger by two findings. The first is that bystander affiliation is more likely to occur if the fight was intense and so likely to elicit greater distress. The second is that the bystander's affiliation is more likely to be directed to an individual victim if the two have a valuable existing relationship. Empathic responses are clearer when distress is unambiguous and a clear relationship has been established. I am more likely to have formed an emotional attachment with my wife than with a stranger in the street, so my distress at her pain will be greater than anything I feel towards a stranger.

Consolation

Empathy can be studied in birds using consolation after a fight as a method. An aggressor attacks a victim while its partner (a bystander) watches. After the fight, either the bystander approaches the victim and comforts it, using a special behavior called bill-twining (bottom left) or the victim solicits comfort from the bystander, who then consoles the victim (bottom right).

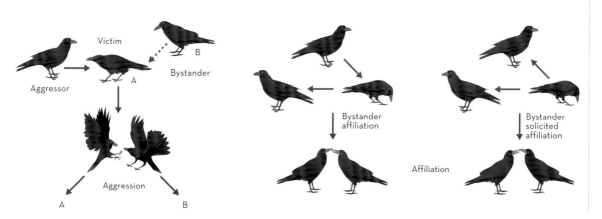

7 NO LONGER BIRDBRAINS

Einsteins of the air

I hope I have convinced you by now that we have underestimated the intelligence of birds, and our reassessment of their capabilities has revealed why some birds should be considered as smart as apes and dolphins, and "no longer birdbrains." Indeed, corvids and parrots may even equal human infants or our earliest ancestors in cognitive prowess. Not surprisingly, it will probably be abhorrent to some to be told that birds may be as smart as humans, especially when birds still dominate our dinner tables or are shot routinely by farmers or hunters.

A TURING TEST FOR BIRDS

Alan Turing, breaker of the Enigma code and inventor of the modern computer, suggested it might be possible to create a machine that is indistinguishable from a human. He devised a test in which a judge had to determine whether two competitors were human or machine based on their answers to a series of questions. To date, there has been little success in finding a computer that can sufficiently mimic a human mind. What about animals? I have described cases in this book attesting to human-like abilities in birds, but perhaps we need to reassess what we mean by human cognition. For example, humans have episodic memories, but those memories are highly influenced by our relationships, other events in our lives, our emotional state, and stories we've read or seen on TV, as well as being embellished by our imagination. However, these influences are probably absent in the episodic memories of animals. They don't have this cultural baggage to influence the purity of their memories. So I see no reason why a bird remembering a specific past event, and even seeing itself taking part in that event in its imagination, should challenge the uniqueness of human episodic memory.

HOW TO MAKE A PIGEON LOOK LIKE AN APE

According to behaviorists such as Robert Epstein, there is no such thing as cognition, because all behavior, including human behavior, is the result of trial-and-error learning or instrumental conditioning.

Left The two species of birds with the biggest brains, such as corvids (left) and parrots, have been demonstrated to have cognitive abilities on a par with the great apes, and in some cases, even humans. This idea is contrary to many people's opinions about animal intelligence, especially birdbrains.

We learn by trying things out and seeing what happens; we don't make predictions by developing mental models. If I want to know how to use a fork, I need to pick it up and see what it can do rather than model its function in my head based on similarities with other, similar tools I've used in the past. Although this idea has been disputed for human behavior, it still holds sway for animal behavior, largely because we only have their behavior to work on. An animal that takes hundreds of attempts to solve a problem is likely to be doing so using trial and error, whereas an animal that learns in only one attempt may be doing it another way.

Epstein used instrumental conditioning techniques to simulate a series of experiments designed to discover whether chimpanzees could communicate using symbols representing words (a core component of human language), solve problems using insight, imitate, or are self-aware. Rather than apply techniques to a bird possibly capable of solving the tasks naturally, they chose pigeons they could train to simulate the complex-looking behaviors of chimpanzees. Epstein suggested that all complex behaviors are the result of linking together simpler components previously rewarded using trial and error. And so, the Columban simulation project was born (*Columba livia* is the Latin name for domestic pigeon). Epstein was able to train pigeons to peck at a dot placed on their chest which could only be seen in a mirror; push a small box directly under a banana hanging overhead, stand on the box, and peck the banana; and communicate information about colors to a neighbor using symbols representing those colors. Each demonstration was the result of meticulous training, yet the researchers claimed that the pigeons' end behavior was the same as the chimps', who could just as easily have learned individual components of each task before displaying them linked together for the final result.

PIGEONS LOVE PICASSO

How do birds categorize stimuli and how do they form concepts? Groups of similar objects can be classified based on shared perceptual features, whereas other objects may share features based on function. Faces can be classified perceptually, as they all have two eyes, two ears, a nose, and a mouth in a specific configuration. Tools are classified by shared functional features—objects do a job that cannot be performed by the body alone—but which can look completely different. Pigeons can be trained to classify objects such as faces but also images that are more conceptual, such as trees, people, or even artistic style. For example, pigeons were trained to group images based on whether they contained a tree, even though the trees were all different species. The birds would differentiate trees from bushes and other vegetation, suggesting they had formed a "tree concept." Strikingly, pigeons could also discriminate pictures by Picasso from those of Monet and, moreover, were able to transfer these painting styles to novel pictures they had never seen before. The pigeons also generalized Monet's painting style to Cézanne and Renoir, and Picasso's painting style to Braque and Matisse!

ANALOGICAL REASONING

Birds also appear capable of forming more complex relational concepts, such as deciding whether two objects are the same or different, or whether one is bigger or heavier than the other. This requires a deeper conceptual understanding than discriminating between perceptual features, as an object has multiple features at the same time. Object B can be smaller than object A, and larger than object C, depending on which object it is being compared to. One concept that has been difficult to examine in animals is the relation between relations, or analogical reasoning. Here, the relationship between pairs of objects is compared. For example, a sample pair A is a large green circle and a smaller yellow circle. Pair B is a large red star and a smaller gold star, and Pair C is a small blue star and a small orange square. Which pair is analogous with Pair A? Answer: Pair B because one of the two objects in the pair is larger than the other, whereas the two objects in Pair C are the same size. If a new Pair

D is added as a sample, with a large blue circle and a small orange star, which pair would now be analogous to Pair D? Now, Pair C would be analogous, but in terms of color, not size, as Pair C matches the sample in color. Until very recently, it was thought that animals could not discriminate objects based on analogies (with the possible exception of a language-trained chimp). However, hooded crows have been shown spontaneously to display analogical reasoning to stimuli based on color, shape, and number.

NUMBER SENSE

Why would a bird need to understand number? Any bird that can distinguish quantities will have an advantage over other birds. There is even evidence that some birds discriminate numbers of objects, such as New Zealand robins counting the number of caches they have made or cuckoos counting the number of eggs they have parasitized. Otto Koehler tested ravens, jackdaws, crows, and various parrots on their ability to count, presenting birds with cards with various sizes of dots on them. The number of dots on one card was greater than the other, yet the surface area of the combined dots was the same on each card. The birds accurately chose the card with the greater number of dots, the cardinal number, up to 6 or 7. Koehler described how the birds could have performed these discriminations. For instance, jackdaws nodded their heads the same number of times as the number of dots, akin to children counting on their fingers. Such abilities are not restricted to adult birds. Even newly hatched domestic chicks have a simple form of arithmetic. Chicks were reared with five identical objects, then, during tests either two or three of the objects disappeared behind one of two opaque screens (either simultaneously or in series). The young chicks inspected the opaque screen hiding the larger number of objects, thereby discriminating between two and three objects. Young chicks therefore appear to have an innate ability to discriminate numbers of objects, although it's not clear why they would need this ability at such a young age.

Number Sense in New Zealand Robins

New Zealand robins, like many birds, are capable of discriminating between low and high quantities of food, however they are also able to count and perform rudimentary arithmetic (addition and subtraction).

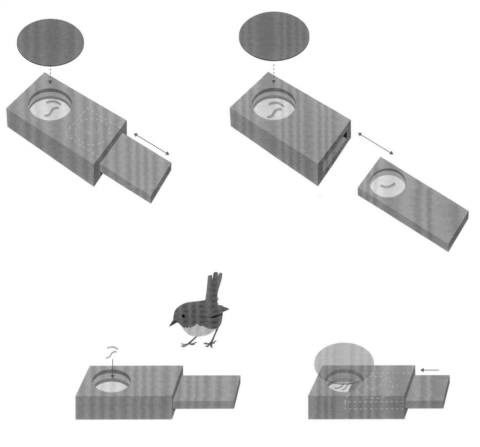

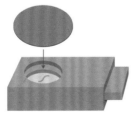

1 An apparatus that can present a mealworm to a robin, but that also conceals a mechanism allowing the worm to seemingly disappear when manipulated by the experimenter was used to test robin's sense of arithmetic in experiments performed in the wild.

2 A robin was shown two worms being placed into the apparatus. A cover was placed over the opening. Unbeknownst to the bird, a secret mechanism inside the box enabled the experimenter to hide the view of the two worms, only revealing one worm.

3 The robin was then presented with the apparatus, and searched for the worms by displacing the covering lid. To its surprise, it discovered that one of the worms had disappeared, violating its expectancy.

4 If the robin understood that the number of worms did not match the number they had seen being placed into the box, or in cases of arithmetic, either being added or taken away from the box, they spent longer searching around the box, than when their expectancy had not been violated (the number was correct based on arithmetic).

AVIAN INNOVATORS

One difficulty in investigating the evolution of intelligence is that we don't have a good proxy for intelligence in the fossil record. We have to rely on other measures—such as brain size, group size, diet, and habitat—that may relate to the need for intelligence in order to survive.

NECESSITY IS THE MOTHER AND FATHER OF INVENTION

A number of candidates have been proposed as driving forces for the evolution of human intelligence: hunting, tool use, culture, sociality, and Machiavellianism. In these cases, intelligence relates to a single domain of knowledge (social or physical), whereas we now know that human (and possibly animal) intelligence spans different domains. The very notion of intelligence is that skills are adapted across contexts for which they didn't evolve. One suggestion that cuts across domain boundaries—thus possibly reflecting a better correlate for intelligence—is innovation. Innovative behavior has helped us to become the dominant species on the planet, allowing us to visualize the smallest objects, communicate across time and space, and even travel to the moon. All from the humble beginnings of a simple stone ax.

INVENTIVE BRAINS IN ACTION

Do birds have brains that can generate new ideas? In order to find out, researchers needed a clear method for assessing claims of innovation across a wide number of birds. Louis Lefebvre and colleagues found the avian research literature was full of short case studies by professional ornithologists, as well as amateur birdwatchers, reporting details of unusual behaviors. These were either novel behaviors or cases of eating a novel food. By trawling through these reports, the researchers built a database of avian innovations, divided up based on bird family. Some families (corvids, birds of prey, and gulls) displayed more innovations than other birds, but surprisingly not parrots, although this is likely to be due to difficulties in observing parrot behavior in the wild. Examples of innovations in corvids far outstripped those of other bird groups. This is not surprising. As I write these words, I can see a jackdaw trying to find its way into a bird feeder in our garden, and with a little more time and effort, it should find an inventive solution to the problem—invention in action.

WHEN EATING VOMIT CAN BE CLEVER

Many of the innovations were reports of eating novel foods. Not surprisingly, those species that were omnivorous incorporated more novel foods into their diet than other birds, especially in harsh conditions when normal foods were unavailable. One of the strangest reported cases was a rook eating frozen vomit! There is also a strong relationship between birds that use tools and those that are inventive, as well as to performance on reversal learning tasks, problem-solving, and social learning. Like primates, innovative birds tend to have the largest brains (as well as the largest mesopallium and nidopallium). However, this remains a correlation, not necessarily a causal relationship. It says little about whether large brains are needed to produce innovations or whether they are the result of innovations (for example, does procuring a better diet result in a bigger brain?).

Above Some parrots have become a common sight in many towns and cities. These rapidly invasive species, are highly adaptable and smart, exploiting many environments to which they are not naturally accustomed.

Right As opportunistic, omnivorous, avian innovators, seagulls have exploited a resource that is high in calories, and a substitute for their natural diet of fish that are dwindling in numbers, and difficult to catch.

OUR CHANGING RELATIONSHIP WITH BIRDS

A consequence of being more innovative is the opportunity it provides for exploiting new habitats, especially when paired with a more explorative personality willing to take risks. Environments are not fixed, so resources can dwindle, especially if climate changes. This may have devastating effects on species that cannot adapt.

ALIEN INVADERS

Consider an individual rigidly adhering to a simple diet, compared to one that experiments with new foods and can move into new areas. If the habitat of these two individuals changes so much that food levels cannot sustain them, only the innovators will survive and produce healthy offspring. For example, birds successfully introduced into new habitats across the world, such as in New Zealand, were found to have large brains and a great innovative tendency, and the ability to adapt to new circumstances may have driven their success in colonization.

STREET SMARTS

A rook sits on a garbage bin in the car park of a service station and pushes its head into the bin pulling out a discarded slice of pizza. However, the bird doesn't stop there, it grabs hold of the liner inside, pulls up one side, places the plastic under its foot, and reaches down to pull up another piece. It repeats this until it has reached the bottom and pulls it up over the side, spilling the contents onto the floor. The rook's partner can also now enjoy the high-calorie contents of the bin.

URBAN WARRIORS

As birds encroach into human environments because of the threat of climate change and habitat destruction, with the subsequent effects of reducing their food supplies and potential nesting sites, we have to adapt to their presence, but they also have to adapt to these strange new worlds. Those birds that are smart enough to adapt their behavior to a changing world will be more successful than those that cannot. As birds start to share our world, they will lose their fear of us and increase their tenacity. Some species, such as gulls, are behaviorally flexible, exploiting a wide variety of resources that put them potentially into conflict with humans. You may recall the story of a gull in Aberdeen, Scotland, who ventured into a local store, picked a certain orange-colored brand of tortilla chips off the shelf, took the packet outside, broke it open, and ate the contents. No other brand was good enough for that gull. Local residents were so enamored that they paid the shop owner for the gull's treats. A consequence of birds entering our world is that they will exploit it. As any gardener will attest, it can be a constant battle preventing birds eating newly planted seeds, and those with water features will have experienced herons swooping down to take precious fish from their ponds. But there are also positive consequences of interacting with birds. In one case, a young girl fed crows in Seattle every day, and in return they brought her a different object on each occasion, possibly as a gift.

Songbirds face problems of increased noise and light pollution when moving from the country into the city. Birds have been able to adapt by changing the duration and frequency of their songs depending on the background noise. Great tits, for example, use higher-frequency singing when the background noise is increased (which is typically of a lower frequency). Black-capped chickadees use shorter, higher-frequency songs when traffic noise is louder, and longer, lower-frequency songs when it has been dampened. Light pollution can also have physiological effects, especially when photoperiod (day length) controls the timing of breeding, nesting, and migration. For example, in towns with street lighting, American robins have shifted to singing earlier in the day if close to artificial lighting.

Right Innovative and adaptable bird species have started to encroach into our worlds as their habitats get smaller, and opportunities for foraging are reduced. Those that can exploit new foods thrive in new environments, but gain the chagrin of those whose environments they invade.

UNIQUELY HUMAN?

The Reverend Henry Ward Beecher once proclaimed, "If men had wings and bore black feathers, few of them would be clever enough to be crows." It's not clear why he wrote this, although it probably wasn't because he was a student of avian intelligence. It's more likely because he was a supporter of the abolition of slavery and was commenting on man's stupidity rather than the cleverness of crows. Whatever his intentions, Beecher was correct and could have included birds with blue, green, red, yellow, and brown feathers as well.

FEATHERED FEATS

Some birds display great feats of memory, recalling the location of thousands of different items, even after long intervals. They travel some of the longest unaided distances of any animals on the planet.

Birds communicate their intentions using visual signals and recognize what others are looking at, even when hidden from view. Their vocal communication shares traits with human language. Birds are social, yet the pair bond is at the heart of their society. Birds form intense long-lasting relationships with others and can remember friends and enemies. They cooperate, share food to curry favors, and help and support one another. Some birds use different tools for different jobs, and stick to the same tool that their peers use, displaying something akin to culture. They create tools to solve novel problems, perhaps using insight. Birds

remember specific events in their past, what happened, where, and when, and use these memories to plan for their future. Because some birds respond to hidden marks on their bodies in the presence of a mirror, they may even be self-aware. Finally, some birds may be able to use another's behavior to predict their intentions, and distinguish between different states of knowledge. Importantly, you don't need to travel to the rainforest or the deep oceans to see these examples of intelligence. All you need is to do is to look out your window, and focus on the activity on your bird table. This list of cognitive achievements suggests that the intellectual abilities of some birds may have been underestimated by the general public. Moreover, birds may even be better models for how human cognition evolved than our ape cousins.

A WORD OF CAUTION

Although I have tried to be cautious throughout this book, it is difficult to go into all the relevant arguments or provide all the essential details of experimental procedures in such a short space. Some of the details and arguments are also quite technical, dependent on understanding experimental design and statistics that would be boring to read and dull to write. However, in all the studies I have described in this book, I have to raise a caveat about interpreting mechanisms underlying behavior. Experiments are designed in an attempt to control for alternative explanations, but scientists frequently disagree about results—either because they have taken a particular position they wish to support or because the data is ambiguous or because of some long-held dogma that is difficult to overturn. In animal cognition, arguments have traditionally been over whether a behavior is the result of learning or cognition. For example, the behavior may be due to a series of experiences linking actions with their consequence (learning) or cognition (transferring a learned rule to a different context or forming a mental image in the imagination). In fact, we now think that both are involved to a greater or lesser degree. How these processes interact will be the focus of future studies.

Left Although many of us are happy to feed garden birds, and enjoy the daily dramas that unfold on our bird tables, not many of us think about the minds of those birds, such as these tree sparrows. Indeed, we probably also enjoy visiting zoos to spy on our primate cousins without realizing that some of our feathered visitors are just as clever and interesting, and only require us to look out our kitchen windows.

Right Birds are all around us, yet most us don't really seem to notice them. We certainly don't appreciate them as social, sophisticated, intelligent beings with the capacity for complex thought. Hopefully, by spending some time to look at what they do, we will start to see some extraordinary similarities with ourselves.

WHEN DOVES CRY

Another important aspect of an animal's psychological makeup is whether they experience emotions. Attributing emotions to animals is fraught with as many problems as attributing cognitive abilities. Some of the lowest points in human history have been due to our negative emotions: hate, fear, greed, grief, and jealousy. Is it the case that birds share these emotions?

WHAT ARE EMOTIONS GOOD FOR?

One problem is that animals look like they have feelings, but this is anthropomorphic. If a dog opens its mouth wide, chases, and wags its tail, we automatically believe the dog is experiencing joy. This isn't very satisfactory from a scientific standpoint, but neither is the killjoy view that the dog's behavior is just a response to reward. Where is the common ground between these opposing positions? Even scientists find it difficult not to attribute emotions to pets. Perhaps, the best place to start is to focus on the simplest emotions, such as fear.

FIGHT OR FLIGHT?

Fear is the emotion most intensively studied in animals, and for which we have the strongest evidence. Fear has a well-defined neural basis focused on the amygdala. Although birds have equivalent brain structures to the amygdala, namely the arcopallium, extended amygdala, and nucleus taenia, few studies have focused on fear in birds. Fear is an adaptation that protects us from external threats, but may not require much thinking. An animal may switch to a state of fear after perceiving an object in its "fear database" (either innate or learned through experience) rather than having to think through various options concerning how to respond. The fearful animal will either avoid interacting with that object, move away from it, or warn others about its presence. The aversive object elicits a fear response that changes the animal's behavior, but we don't know whether or not the animal feels fear. We can only extrapolate from its behavior, or we can study its brain activity. A recent study in crows found that neural circuits equivalent to the mammalian emotional brain were activated after perceiving threatening faces.

FORTUNE FAVORS THE SUBORDINATE

Birds have a natural tendency to be wary of novel objects (neophobia), but there will always be some birds that will investigate new objects. These pioneers tend to be low-ranking, benefiting from overcoming their fears because otherwise they would never access the best resources. If their attempt at contact is successful, it may spread through a population, and the new object may no longer be avoided or may even be incorporated into the group's food database. Some stimuli, such as predators, are in the fear database from birth, being so dangerous that there is no leeway for learning. A bird's response to fear depends on its personality, such as whether it is more or less likely to approach a novel object or enter a new environment, or whether it is more or less likely to explore or stay on the sidelines. These tendencies lie on the shy–bold continuum, with the shy more wary and fearful and the bold more likely to take risks.

LOVE AND LOSS

How could we tell if a bird experienced grief? Pet birds separated from their owners, or lifelong monogamous birds whose partner dies, look as if they are grieving for their loss. Bonded birds spend all their time together, preening, sharing food, and rubbing beaks—all behaviors used to reduce stress. Are these expressions of love or simply physiological processes that keep a pair together to raise healthy offspring? It looks like love because birds act like us in the same context. Similar hormones—mesotocin (the avian version of oxytocin) and arginine vasopressin—are involved in love and bonding, but that doesn't mean the experiences are the same. If they do experience love, do they also experience the emotional pain when their loved one is separated from them? Paired birds separated from their partner stop eating and self-preening, they look and continue calling for them, and are listless and drooped. However, we still cannot say that the birds experience grief.

Right Many birds form strong bonds with both their partner, but also their chicks. Similar hormones to those implicated in human love appear to be involved in these bonding processes in birds.

Birds just wanna have fun

Play makes us happy. It is difficult not to think that birds feel the same when we observe crows sliding down snowy rooftops or swans surfing along the crest of a wave. Birds' play and their emotional brains are similar to ours, so do they experience the same positive emotions as us?

WANTING AND LIKING

The mammalian brain contains networks for positive reward: the wanting system and the liking system. The wanting system causes an animal to seek out something rewarding, whereas the liking system provides a sense of pleasure associated with the reward. The wanting system drives an animal to continue playing, and the liking system provides pleasure experienced during play. Both systems are dependent on the distribution of dopamine. Dopamine is a neurotransmitter found throughout the mammalian brain, and in comparable regions of the avian brain such as the nidopallium, mesopallium, and even the song control system. Endogenous opiates are also essential for reward and are found in similar regions to dopamine. Birds therefore have brains set up for experiencing similar emotions to mammals, including humans. Play may be one route by which they experience these emotions.

PLAY ON THE BRAIN

Play has been most intensely described in birds and mammals. As there are only few examples of play in captive reptiles, it is likely that play evolved independently, especially when the most complex forms of play are only found in the most intelligent species. Play may be dependent on cognition, especially when species with relatively larger brains tend to play more. Play is more common in species with a longer development, giving young animals the opportunity to learn about how the world works. Cases of play in birds are relatively uncommon, with only 1 percent of the approximately 9,000 species displaying one of the three types of play. Those cases tend to be in crows and parrots, which typically play

Left Kea are parrots found in the mountainous regions of New Zealand's South Island. They are known as the clowns of the bird world, as they seem to relish destroying human property. Is this really play, or are they just investigating something new?

like primates and carnivores, the two mammalian groups with the highest incidences of play. Examples included acrobatics, playing with objects and play between conspecifics, including play fighting. Crows and parrots have evolved specialized play signals to help aid their play partners in distinguishing cases of play from real aggression.

PLAY FORMS

Birds display three types of play. First, locomotor play, which includes aerial acrobatics, hanging, and flying upside down. Ravens and raptors consistently display all sorts of acrobatic acts while flying. Second, object play involves the close inspection of objects to learn about how they work and whether they are edible. Some captive birds have taken object play to the next level, using novel objects as tools. Such birds

have to approach and manipulate objects not usually encountered in their natural environment, investigate them, then discover whether they can be used as tools. For example, keas are attracted to unfamiliar objects, and are notorious for destroying the external fixtures on cars, raiding rubbish bins on campsites, and so on. It is difficult not to see their wanton destruction as them having fun. Finally, social play provides a method for learning how to fight and court, involving chasing, tussling, and rough and tumble. Social play may utilize objects, where favored objects are stolen or fought over. For example, captive rooks play tug-of-war with strips of newspaper, even when the birds are standing in thousands of paper strips that look exactly the same as the piece they are fighting over. This strongly suggests that the birds are having fun rather than needing that specific piece of paper.

Play on the Brain

The distribution of dopamine neurons and receptors in the raven brain may play a role in different forms of play.

- k opiate receptors
- dopamine receptors
- m opiate receptors
- → dopamine projections
- song system

Human language and birdsong

We have already addressed the idea that three families of birds—songbirds, parrots, and hummingbirds—are relatively rare in the animal kingdom in learning their vocalizations, and that they have developed similarly structured neural circuits to perform this task. Perhaps surprisingly, humans learn language via a similar process to birds, namely, during a sensitive period outside of which language doesn't develop properly.

BIRDSONG AS A MODEL FOR HUMAN LANGUAGE

Human vocal learning and production is dependent on similar neural pathways to birds. Auditory information (words) is processed by the auditory speech areas of the temporal cortex (Wernicke's area), learned, and turned into speech by Broca's area in the frontal cortex. A projection from the face motor area controls the physical act of speaking. Birdsong provides a suitable animal model for human language despite differences or absences of processes in birds; because songs are learned, elaborate patterns of vocalizations that have a specific structure or syntax impute meaning to the vocalization in a way that is not present to the same degree in calls. Human language and birdsong also develop during two phases: auditory learning and sensory-motor vocal learning.

RECURSION IS RECURSION, IS RECURSION, IS RECURSION . . .

One area in which birds and humans differ is the potential complexity of their utterances. Bird songs are syntactically structured, in terms of the ordering of phrases and notes in a sequence, but they are not capable of recursion—an embedding of phrases inside other phrases to create hierarchical sequences of potentially infinite length. My favorite example of a

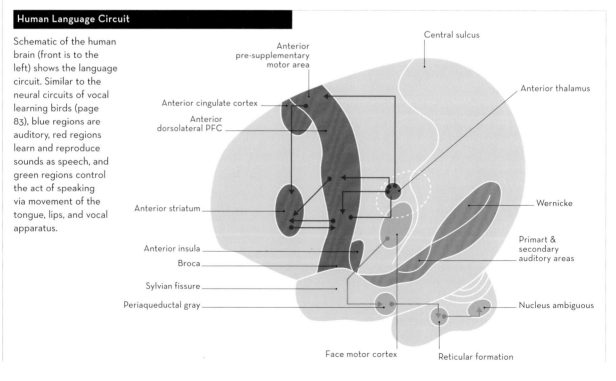

Human Language Circuit

Schematic of the human brain (front is to the left) shows the language circuit. Similar to the neural circuits of vocal learning birds (page 83), blue regions are auditory, red regions learn and reproduce sounds as speech, and green regions control the act of speaking via movement of the tongue, lips, and vocal apparatus.

Central sulcus

Anterior pre-supplementary motor area

Anterior cingulate cortex

Anterior dorsolateral PFC

Anterior thalamus

Anterior striatum

Wernicke

Anterior insula

Broca

Primart & secondary auditory areas

Sylvian fissure

Periaqueductal gray

Nucleus ambiguous

Face motor cortex

Reticular formation

simple recursion is a common sign in the UK, "Sign not in use." The sign actually refers to another (electronic) sign that isn't working, yet the fact that one sign is obviously in use yet says that it is not working sets up an infinite loop—how can a sign be both working and not working?. This is recursive because one piece of information is embedded in another that is higher in the structural hierarchy. In human language, an example of recursion is, "Harry knew that [Susan thought that [Jim believed that]] Bob was going to town." The important information is, "Harry knew that Bob was going to town," but additional information embedded in this could be relevant for social perspective, without causing much distortion to the overall meaning of the sentence. Despite some gallant attempts to determine whether birds, such as starlings and zebra finches, understand recursive structure in song patterns, the evidence so far argues against this particular trait as playing a role in song learning.

FOXY GENES

Human speech is under both genetic and environmental control (nature and nurture). Although there is no 1:1 relationship between a gene and a behavior, mutations of some genes do have effects on a specific behavior.

In the case of language, mutations of the gene encoding the transcription factor FOXP2 cause a specific speech disorder in three generations of one family. This disorder, called developmental verbal dyspraxia (DVD), causes problems with the movement of lips, tongue, jaw, and palate, which all contribute to normal speech production. The FOXP2 gene is evolutionarily conserved, as it is found across a wide range of species, and appears to be important for neural development (building a brain). It is found in the songbird brain, expressed primarily in Area X of the striatum, part of the vocal imitation loop. FOXP2 appears to play a role in vocal plasticity, as more FOXP2 gene expression is seen in Area X when song becomes unstable (before it becomes stereotyped and crystallized). Birds with inactive FOXP2 genes (knocked out) during the sensorimotor song-learning phase failed to copy tutor songs properly, with many inaccuracies in the produced song. This is akin to the word production problems displayed by DVD patients.

Below Zebra finches do not sing the most melodious song in the natural world, but the fact that they chirp very simple songs has made them ideal models for studying the neural basis of bird song, and whether it shares any properties of human language.

Evolving intelligence

Until the last century, we humans assumed we were alone in our intelligence and did not consider ourselves a part of the animal kingdom. Many are still resistant to the idea that we descended from an ancestor held in common with chimpanzees, and we still think of ourselves as top of the class when it comes to smarts.

THE GREAT CHAIN OF BEING

Before Charles Darwin, humans and animals were compared on what the Greek philosopher Aristotle called the Great Chain of Being or scala naturae. Animals were positioned on a ladder, progressing from invertebrates on the bottom rung; to fish and amphibians on the next rungs up; then to reptiles, birds, and, finally, mammals, and all the way to humans at the top. This misplaced idea had a detrimental effect on our understanding of avian intelligence.

EVOLVING BODIES, EVOLVING MINDS

Now we know that evolution affects animal minds in the same way as animal bodies. With bodies, closely related species look similar because they have evolved the same physical features from a common ancestor.

Two species of crows look the same because they shared a common ancestor with those features and because evolution is conservative, keeping features that work well and maintain reproductive fitness (producing sufficient numbers of healthy offspring). Most crows have black feathers, but some may have evolved a slightly different-shaped beak to adapt to a different type of diet—seeds, fish, carrion, insects—or even one that is multifunctional and can be used to make tools. Does evolution work in a similar way for birds' minds?

HOMOLOGIES & ANALOGIES

Shared features with a common ancestor are called homologies. Features are homologous if the simplest explanation for their evolution is that they were

derived from a common ancestor. However, there are cases of species sharing traits that cannot be derived from a common ancestor. Sharks and dolphins have evolved a similar streamlined body shape enabling them to swim very quickly, probably as a method for chasing prey. Pterosaurs, insects, bats, and birds all evolved appendages for flight, but not from a common ancestor or else more species would be able to fly. Shared features based on functional similarities that cannot be due to shared ancestry are called analogies. Features are analogous if the two species are distantly related, and most other sister species do not possess the same features. Therefore, the simplest explanation for their presence is convergence onto a similar solution to an environmental challenge, even if the feature is structurally different in the two species. Complex eyes evolved many times in order to process more visual details, such as color, than was possible with a simple visual receptor tuned to perceiving light versus dark.

CONVERGENT EVOLUTION OF INTELLIGENCE

As behavior also evolves, distantly related animals displaying similar cognitive abilities, but with differently structured brains, are likely to have evolved those same intellectual skills independently. Corvids and apes face many of the same socio-ecological challenges and potentially use similar cognitive solutions to overcome them. For example, corvids and apes live in large complex social groups in which individuals are at an advantage if they can determine where, what, and why another is looking at something. Crows and chimpanzees can make sophisticated mental calculations of another's perspective based on the angle of their eye gaze and use this information accordingly to compete over food. These comparable abilities cannot be the result of homology, as the last common ancestor to apes and crows lived more than 300 million years ago, and most birds and mammals do not use similar abilities in their social interactions. It isn't clear how these two distantly related groups have homed in on the same solution to similar problems. However, the fact that corvids and apes share a number of biological and psychological traits that should assist in their information processing (brains evolved to process lots of information quickly, color vision, manual dexterity, sophisticated communication system, omnivorous diet, bonded societies) may be one factor behind their shared skills.

Below Corvids and parrots are not alone in their expression of intelligent behaviour. Many creatures, such as elephants, dolphins, hyenas, raccoons, apes and monkeys that share certain biological traits, such as large brains and complexity sociality, are also members of the Clever Club, having evolving their intelligence independently.

Welcome back to the Clever Club

By assessing the traits shared by corvids and apes that are absent in closely related groups, such as non-corvid songbirds or lemurs, we can begin searching for other potential members of the Clever Club. So far, we can probably include parrots, elephants, whales, and dolphins, and possibly some monkeys, hyenas, and raccoons, as well as certain members of the weasel family.

Before we can be sure of club membership, we need to start examining the cognitive capacities of a wider range of animals, designing tests that may be applied to animals that do not rely on the same senses or need to manipulate objects without the aid of hand or beak. Dolphins, for example, do not have limbs with which they can build tools, and thus they rely on their auditory system to learn about the world, rather than the predominant sense of vision used by land-based mammals and birds. Dolphins fail cognitive tasks requiring vision but succeed on auditory versions of the same tasks.

CLEVER CLUB, CLEVER BRAINS

At the beginning of this book, I considered how the avian brain could possibly process information as efficiently as the mammalian brain. The corvid and parrot brains are as large as great ape brains relative

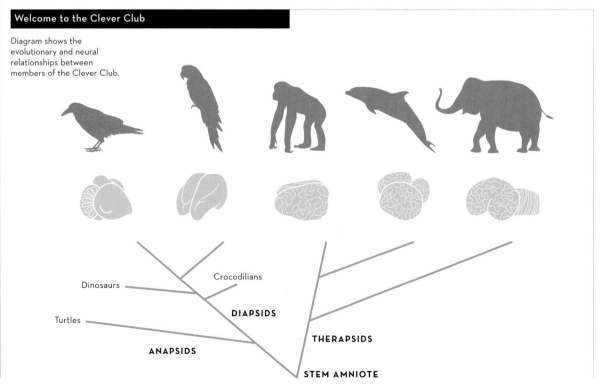

Welcome to the Clever Club

Diagram shows the evolutionary and neural relationships between members of the Clever Club.

Dinosaurs

Crocodilians

DIAPSIDS

Turtles

THERAPSIDS

ANAPSIDS

STEM AMNIOTE

to the size of the animal. The Clever Club of mammals—apes, dolphins, whales, and elephants—all have very large brains. Humans have the largest brains for their body size of any creature, as well as a larger prefrontal cortex. One small detail that may set these brains apart from other animals is von Economo neurons or spindle cells. These neurons are only found in the anterior cingulate cortex and the insula, two regions involved in the socio-emotional processing of humans, apes, dolphins, whales, and elephants. Structurally, they are quite different from the more profuse cortical neurons, called pyramidal cells, which have a long dendrite extending upward, and an array of dendrites radiating out from the cell body, By contrast, spindle cells have single dendrites extending upward and downward from the cell body, with an absence of the dendritic array seen in pyramidal cells. Currently, the function of spindle cells is the subject of speculation, although the most popular suggestion is a role in social cognition. Perhaps they are capable of efficiently transmitting information across longer distances than is possible with pyramidal cells alone.

As only mammalian members of the Clever Club have these specialized neurons, they may aid in transmitting information across the huge distances in these larger brains, thereby increasing their efficiency.

DO BIRDS HAVE SPINDLE CELLS?

As far as I am aware no one has yet looked into this. However, it is more likely that the avian members of the Clever Club—the crows and parrots—possess their own special neurons not found in other birds, rather than these mammalian-specific cell types. However, if spindle cells do increase the efficiency of information transfer across larger brains, then the nucleated structure of the avian brain may already be an evolved solution to the problem of efficient information transfer. It would be quite something if the "best" mammalian brains evolved a cell type that made them more like avian brains!

Below Jackdaws, like other corvids, have very large brains for their body size. Indeed, they are relatively the same size as chimpanzees, and they have to have a low body weight to fly. Perhaps birds have evolved certain efficiency tricks to keep their weight down while increasing their brain size?

A cognitive toolkit

Do we have any idea how both corvids and parrots have achieved cognitive feats that would make a bonobo proud? In a survey of the cognitive abilities of apes and corvids, four cognitive tools were proposed as underlying the complex cognition of these two groups—tools not found in closely related species.

Above Birds, such as blue jays, have to be highly flexible in their behavior in order to cope with a rapidly changing environment. Possessing a cognitive toolkit, rather than relying on trial-and-error learning can put them ahead of the competition when the going gets tough.

It is worth reiterating here the difference between cognition and intelligence. Many animals use cognition in their daily lives. For example, bees use a sophisticated "dance language" to transmit information to other hive members about the location of a food source that cannot be seen from the hive. Monkeys trade objects for foods of differing value, and domestic chickens just a few days old display mathematic skills that rival any human toddler. Yet, these species do not necessarily possess intelligence, the ability to adapt knowledge and skills, because such knowledge and skills are tied to the context in which they were learned or for which they evolved. They are not flexible. Flexibility is the first of our cognitive tools that allows crows and apes to switch or update strategies when conditions change. For example, it may be chilly when food is cached, but temperature may suddenly increase and affect the length of time that the food remains fresh. A flexible bird will be able to update its memory based on this new information in order to recover the food earlier than originally planned. The second tool is imagination. It has been previously suggested this is a uniquely human attribute, and its mooted presence in any animal is controversial, yet there is supporting evidence that some birds and apes

Cognitive Toolkit

Complex cognition in corvids, apes, and probably other members of the Clever Club consists of at least four cognitive tools: causal reasoning, flexibility, prospection, and imagination. These are represented here for corvids. (Upper left) A rook observes the consequences of dropping stones into water. (Upper right) A scrub jay caches different foods in different trays, but needs to update its memory when information about decay rates changes. (Lower left) A scrub jay watches another scrub jay caching; the first bird re-caches the food when the potential thief has left the scene, so as to protect it for later. (Lower right) A rook creates a hook out of wire to retrieve a bucket at the bottom of a tube.

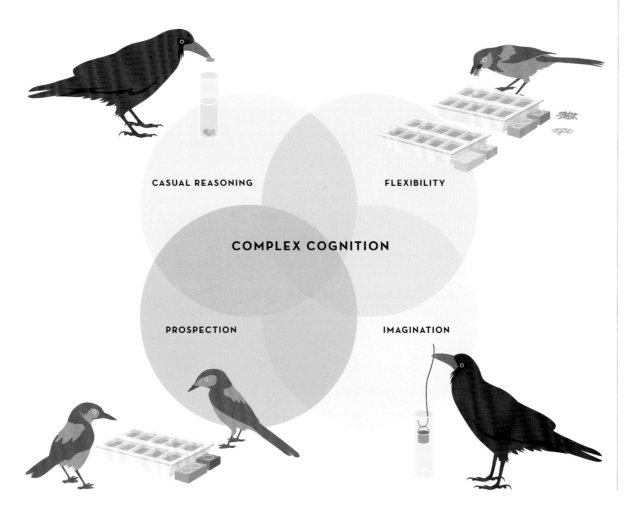

CASUAL REASONING

FLEXIBILITY

COMPLEX COGNITION

PROSPECTION

IMAGINATION

at least use mental trial and error to think through a problem without having to try it out physically first. The case of wire bending is one an imagination would seem a reasonable explanation. Another is choosing which tool to save for use on a puzzle box the next day. Tied to imagination and dependent upon it is prospection: thinking about the future and planning for alternative futures. Prospection requires both an imagination, as the future does not yet exist, and flexibility, as the future is not fixed and can change, so plans need to be flexible. The final

cognitive tool is causal reasoning. This is the ability to understand that certain actions have specific consequences. Predicting the link between action and outcome cannot be seen, but can be learned over repeated presentations. The key test is to transfer what has been learned for one set of stimuli in one context and adapt it to another where similar rules will work. These four tools (and others) may be at the core of corvid and ape intelligence, and possibly that of other members of the Clever Club as well.

THE LANGUAGE OF THE BIRDS

Throughout human history, birds have been seen as mystical or divine, communicating with just a few initiated humans through a special language—the language of angels. Oracles with the ability to interpret this language were deemed capable of predicting the future and considered party to secret information only previously available to the gods.

There are oracles of ancient ingenuity that still persist today—examples of the symbiotic coexistence of the human and avian worlds—which seem capable of reading the behavior of birds in order to enhance human lives: cormorants helping Japanese fisherman catch fish, honey guides leading African bushmen to beehives and the rich honey inside, and ravens leading hunters to wolves feasting on carcasses. But although the modern scientific methods we use to reveal the cognitive secrets of birds would look magical to our ancestors, they don't make us gods, only very lucky.

THE BIRDS ARE THE KEEPERS OF OUR SECRETS

The last twenty years has seen a sea change in how we perceive birds. I hope that the studies I've described in this book give you a flavor of the intellectual capabilities of a tiny number of the 10,000 avian species on Earth. It is time we stopped using the derogatory term "birdbrain." Studies of birds have exposed intimate details of their complex social and emotional lives, but there is so much more to learn about them.

ARE BIRDS CONSCIOUS?

Some scientists are convinced that in discovering the intellectual abilities of animals, we reveal something about their conscious inner nature. They stress that we should treat conscious animals as we would treat another human being. I do not subscribe to this view. Animals are not people, no matter how much they may resemble us physically or behaviorally. That doesn't mean they don't deserve our respect or care, far from it. But, by focusing on whether animals are like us, we do them a disservice in treating them as we would expect to be treated rather than tailoring our attention to the individual needs and motivations that make them unique—the things that may well have attracted us to them in the first place.

Above Are these ravens conscious of the consequences of their actions if they antagonize this wolf? Do they know that by following the wolf, they maybe able to exploit her kill? How would we ever find out?

We now know more about the intelligence of birds than ever before. But this is a very exciting time not only because we know so much, but because we recognize how much more there is to find out. Unfortunately, in this book it has only been possible to dip my toes into the huge variety of research currently being conducted in the field; there is so much more I wanted to cover, but couldn't for want of space. But I hope that having read it, you have a new appreciation for the winged friends with which we share our lives, and that you keep an eye out for the feathered apes in your garden.

Right Cormorants aid Japanese fisherman to catch fish, although by the coercion of their human masters, rather than a feeling of altruism. The fishermen select a team of 10-12 cormorants, who are restrained by ropes attached to their bodies, and rings around their necks to prevent them eating the larger fish they catch. Each cormorant catches up to six fish which they can keep in their throats. They get to eat all the smaller fish they catch.

APPENDICES

GLOSSARY

Adaptive specializations Anatomical, physiological, behavioral and cognitive traits that have evolved to deal with certain ecological demands, such as finding enough food or dealing with social partners. For example, having an impressive spatial memory may be an adaptive specialization for food caching and having to find it later.

Altricial Young organisms that are born incapable of moving around or feeding themselves after birth, requiring a period of care from a parent or guardian. Examples are crows, and parrots.

Analogical reasoning Using analogy (inference that one thing is similar to another) to make judgments about the relationships of other things. Plays an important role in problem-solving.

Autonoetic consciousness The ability to place ourselves mentally in a different time and place, either in the past or the future.

Behavioral flexibility The ability to respond adaptively to changing circumstances or conditions, such as climate or food availability.

Cache protection strategies The abilities of food-storing animals to protect their food stores using distance, light levels, barriers, confusion and other tactics.

Caching The act of storing food by hiding or burying it, either in single sites or as a collection of stores in a larder.

Causal reasoning The ability to identify the relationship between specific events (causes), and the specific effects of those events. For example, presenting pressure to an object may cause it to fall over.

Concepts Generalization across stimuli that differ in appearance, but which share common functional features.

Declarative memory Memory for facts, and personal history (for example, knowledge), rather than memory for doing (such as, habits). Consists of episodic and semantic memory.

Dominance hierarchy Members of a social group interact aggressively, and can be ranked based on their ability to win or lose such aggressive encounters.

Dopamine A neurotransmitter in the brain that plays a role in reward learning and pleasure.

Episodic memory One of two types of declarative memory for specific events in an individual's past that can be tagged with a place, time and content (what, where and when).

Executive functions Cognitive processes, such as working memory, attention, planning, reasoning, flexibility and problem-solving, involved in the management and regulation of behavior.

Experience projection The ability to use personal experience to simulate what another may be thinking.

Extractive foraging The act of locating and processing encased or embedded food, such as roots, tubers, fruit or nuts.

FOXP2 A gene that encodes the forkhead box protein P2 that is involved in brain development and subsequently the development of human speech and language.

Hippocampus An extension of the cortex/pallium that plays an essential role in spatial navigation, as well as short- and long-term memory, and imagination.

Intelligence A collection of domain-general mental abilities to flexibly solve problems outside of the context in which they were adapted.

Neophobia Fear of new objects, places or events.

Neurogenesis The creation of new neurons, particularly in the hippocampus and song control system of birds, either seasonally or to play a role in specific behaviors, such as song or caching.

Neurotransmitter Chemical released by nerve cells that function in communicating between nerve cells.

Oscines Songbirds; members of the Passerine family in which (primarily) males learn their songs from a tutor during a sensitive period.

Pair-bond Strong affinity that develops between a male and a female in a pair, but that can also occur within a same-sex pairing. The bond often leads to offspring, and can last throughout the pair's lives.

Pallium Area of the brain corresponding to layers of white and grey matter covering the upper surface of the cerebrum of the brain. The pallium is involved in sensory processing, memory, emotion, reward learning and decision-making, and found in reptiles, birds and mammals.

Precocial Young organisms that are already mature and mobile from the moment of birth or hatching. They are able to move and find food by themselves. Examples are ducks, chickens and geese.

Prefrontal cortex Area of the mammalian cortex, involved in orchestrating thoughts and actions. In primates, it is located at the frontal part of the brain, separated into orbitofrontal, medial and dorsolateral divisions, each with differing roles in behavior.

Prospection The mental ability to think about and plan for potential futures.

Recursion A potentially infinite loop of information that allows the structure of complex syntactical structures in human language. May be one of the cognitive traits that separate humans from animals.

Scala nature An idea, attributed to Aristotle, that animals may be ranked, from lower to higher, based on their perceived physical (or cognitive) dissimilarity to humans, thereby counter to the ideas of Darwinian evolution.

Selection pressures A cause (perhaps difficulty in finding food or evading a predator) that reduces reproductive success in a population. Inherited traits, such as intelligence, can mitigate the effects of those causes, so driving evolution.

Semantic memory One of two types of declarative memory, referring to general world knowledge (facts, ideas) that is collected throughout an individual's life.

Social intelligence Mental abilities adapted for processing information about the social world.

Sub-oscines Members of the Passerine family, close relatives of songbirds, in which males and females sing, but do not learn their songs from a tutor.

Theory of mind The attribution of mental states, such as beliefs, desires, intentions and knowledge, to another individual in order to understand and predict their behavior.

Transitive inference A form of inferential reasoning that can determine relationships between objects based on their rank order. So, if A > B, B > C and C > D, then we can infer that B > D.

REFERENCES

Akins, CK & Zentall, TR (1996). *Imitative learning in male Japanese quail using the two-action method.* J Comp Psychol, 110, 316-320 (C4).

Auersperg, AMI et al (2011). *Flexibility in problem solving and tool use of kea and New Caledonian crows in a multi access box paradigm.* PLoS ONE, 6, e20231 (C5).

Auersperg, AMI et al (2012). *Spontaneous innovation in tool manufacture and use in a Goffin's cockatoo.* Curr Biol, 22, R1-R2 (C5).

Balda, RP & Kamil, AC (1989). *A comparative study of cache recovery by three corvid species.* Anim Behav, 38, 486-495 (C2).

Beck, SR et al (2011). *Making tools isn't child's play.* Cognition, 119, 301-306 (C5).

Bingman, VP et al (2003). *The homing pigeon hippocampus and space.* Brain Behav Evol, 62, 117-127 (C2).

Bird, CD & Emery, NJ (2009). *Insightful problem solving and creative tool modification by captive nontool-using rooks.* PNAS, 106, 10370-10375 (C5).

Bird, CD & Emery, NJ (2009). *Rooks use stones to raise the water level to reach a floating worm.* Curr Biol, 19, 1410-1414 (C5).

Bugnyar, T (2010). *Knower-guesser differentiation in ravens.* Proc Roy Soc B, 283, 634-640 (C6).

Bugnyar, T & Heinrich, B (2005). *Ravens, Corvus corax, differentiate between knowledgeable and ignorant competitors.* Proc Roy Soc B, 272, 1641-1646 (C6).

Carter, J et al (2008). *Subtle cues of predation risk: starlings respond to a predator's direction of eye gaze.* Proc Roy Soc B, 275, 1709-1715 (C3).

Catchpole, C & Slater, PJ (2008). *Bird Song: Biological themes and variations.* Cambridge University Press: Cambridge, UK (C3).

Cheke, LG et al (2011). *Tool-use and instrumental learning in the Eurasian jay.* Anim Cogn, 14, 441-455 (C5).

Cheke, LG et al (2012). *How do children solve Aesop's Fable?* PLoS ONE, 7, e40574 (C5).

Clayton, NS & Dickinson, A (1998). *Episodic-like memory during cache recovery by scrub jays.* Nature, 395, 272-274 (C2).

Clayton, NS & Emery, NJ (2015). *Avian models for human cognitive neuroscience.* Neuron, 86, 1330-1342 (C1).

Clayton, NS & Krebs, JR (1994). *Hippocampal growth and attrition in birds affected by experience.* PNAS, 91, 7410-7414 (C2).

Colombo, M & Broadbent, N (2000). *Is the avian hippocampus a functional homologue of the mammalian hippocampus?* Neurosci Biobehav Rev, 24, 465-484 (C2).

Cristol, DA et al (1997). *Crows do not use automobiles as nutcrackers.* The Auk, 114, 296-298 (C5).

Curio, E et al (1978). *Cultural transmission of enemy recognition.* Science, 202, 899-901 (C4).

Dally et al (2006). *Food-caching western scrub-jays keep track of who was watching when.* Science, 312, 1662-1665 (C6).

Dally et al (2006). *The behaviour and evolution of cache protection and pilferage.* Anim Behav, 72, 13-23 (C6).

Dally, JM et al (2008). *Social influences on foraging by rooks (Corvus frugilegus).* Behaviour, 145, 1101-1124 (C4).

Dally, JM et al (2010). *Avian theory of mind and counter espionage by food-caching western scrub-jays (Aphelocoma californica).* Eur J Dev Psychol, 7, 17-37.

Diamond, J & Bond, AB (2003). *A comparative analysis of social play in birds.* Behaviour, 140, 1091-1115 (C7).

Emery, NJ (2000). *The eyes have it: the neuroethology, function and evolution of social gaze.* Neurosci Biobehav Rev, 24, 581-604 (C3).

Emery, NJ & Clayton, NS (2001). *Effects of experience and social context on prospective caching strategies by scrub jays.* Nature, 414, 443-446 (C6).

Emery, NJ & Clayton, NS (2004). *The mentality of crows: Convergent evolution of intelligence in corvids and apes.* Science, 306, 1903-1907 (C7).

Emery, NJ & Clayton, NS (2015). *Do birds have the capacity for fun?* Curr Biol, 25, R16-R20 (C7).

Emery, NJ et al (2007). *Cognitive adaptations of social bonding in birds.* Phil Trans Roy Soc B, 362, 489-505 (C4).

Epstein, R et al (1981). *"Self-awareness" in the pigeon.* Science, 212, 695-696 (C6).

Epstein, R et al (1984). *"Insight" in the pigeon.* Nature, 308, 61-62 (C5).

Fisher, J & Hinde, RA (1949). *The opening of milk bottles by birds.* Br Birds, 42, 347-357 (C4).

Flower, TP et al (2014). *Deception by flexible alarm mimicry in an African bird.* Science, 344, 513-516 (C3).

Fraser, ON & Bugnyar, T (2010). *Do ravens show consolation?* PLoS ONE, 5, e10605 (C4).

Fraser, ON & Bugnyar, T (2011). *Ravens reconcile after aggressive conflicts with valuable partners.* PLoS ONE, 6, e18118 (C4).

Frost, BJ & Mouritsen, H (2006). *The neural mechanisms of long distance navigation.* Curr Op Neurobiol, 16, 481-488 (C2).

Garland, A & Low, J (2014). *Addition and subtraction in wild New Zealand robins.* Behav Proc, 109, 103-110 (C7).

Gentner, TQ et al (2006). *Recursive syntactic pattern learning by songbirds.* Nature, 440, 1204-1207 (C7).

Gunturkun, O (2005). *The avian "prefrontal cortex" and cognition.* Curr Op Neurobiol, 15, 686-693 (C1).

Haesler, S et al (2004). *FoxP2 expression in avian vocal learners and non-learners.* J Neurosci, 24, 3164-3175 (C7).

Healy, SD & Hurly, TA (1995). *Spatial memory in rufous hummingbirds.* Anim Learn Behav, 23, 63-68 (C2).

Healy, SD et al (1994). *Development of hippocampus specialization in two species of tit (Parus sp.).* Behav Brain Res, 61, 23-28 (C2).

Henderson, J et al (2006). *Timing in free-living rufous hummingbirds.* Curr Biol, 16, 512-515 (C2).

Herrnstein, RJ et al (1976). *Natural concepts in pigeons.* J Exp Psychol: Anim Behav Proc, 2, 285-302.

Heyers, D et al (2007). *A visual pathway links brain structures active during magnetic compass orientation in migratory birds.* PLoS ONE, 2, e937 (C2).

Hopson, JA (1977). *Relative brain size and behaviour in archosaurian reptiles.* Ann Rev Ecol Sys, 8, 429-448 (C1).

Hunt, GR (1996). *Manufacture & use of hook-tools by New Caledonian crows.* Nature, 379, 249-251 (C5).

Hunt, GR & Gray, RD (2002). *Diversification and cumulative evolution in New Caledonian crow tool manufacture.* Proc Roy Soc B, 270, 867-874 (C5).

Hunt, GR & Gray, RD (2003). *The crafting of hook tools by wild New Caledonian crows.* Proc Roy Soc B: Biol Lett, 271 (S3), S88-S90 (C5).

Hunt, GR & Gray, RD (2004). *Direct observations of pandanus-tool manufacture and use by a New Caledonian crow.* Anim Cogn, 7, 114-120 (C5).

Hurly, TA & Healy, SD (1996). *Memory for flowers in rufous hummingbirds: location or local visual cues?* Anim Behav, 51, 1149-1157 (C2).

Iglesias, TL et al (2012). *Western scrub-jay funerals: cacophonous aggregations in response to dead conspecifics.* Anim Behav, 84, 1103-1111 (C7).

Jarvis, ED (2007). *Neural systems for vocal learning in birds and humans.* J Ornithol, 148 (S1): S35-S44 (C3).

Jarvis, ED et al (2005). *Avian brains and a new understanding of vertebrate brain evolution.* Nat Rev Neurosci, 6, 151-159 (C1).

Jelbert, SA et al (2014). *Using the Aesop's Fable Paradigm to investigate causal understanding of water displacement by New Caledonian crows.* PLoS ONE, 9, e92895 (C5).

Jouventin, P et al (1999). *Finding a parent in a king penguin colony: the acoustic system of individual recognition.* Anim Behav, 57, 1175-1183 (C3).

Kamil, AC & Cheng, K (2001). *Way-finding and landmarks: The multiple bearings hypothesis.* J Exp Biol, 2043, 103-113 (C2).

Karten, HJ & Hodos, W (1967). *A Stereotaxic Atlas of the Brain of the Pigeon (Columba livia).* John Hopkins Press: Baltimore, MD (C1).

Kelley, LA & Endler, JA (2012). *Illusions promote mating success in great bowerbirds.* Science, 335, 335-338 (C3).

Koehler, O (1950). *The ability of birds to "count".* Bull Anim Behav, 9, 41-45 (C7).

Lefebvre, L et al (1997). *Feeding innovations and forebrain size in birds.* Anim Behav, 53, 549-560 (C7).

Lefebvre, L et al (2002). *Tools and brains in birds.* Behaviour, 139, 939-973 (C5).

Levey, DJ et al (2009). *Urban mockingbirds quickly learn to identify individual humans.* PNAS, 106, 8959-8962 (C3).

Liedtke, J et al (2011). *Big brains are not enough: performance of three parrot species in the trap tube paradigm.* Anim Cogn, 14, 143-149 (C5).

Marler, P & Tamura, M (1964). *Culturally transmitted patterns of vocal behaviour in sparrows.* Science, 146, 1483-1486 (C4).

Marzluff, JM et al (2012). *Brain imaging reveals neuronal circuitry underlying the crow's perception of human faces.* PNAS, 109, 15912-15917 (C3).

Nottebohm, F et al (1976). *Central control of song in the canary, Serinus canaries.* J Comp Neurol, 165, 457-486 (C1).

O'Connell, LA & Hofmann, HA (2011). *The vertebrate mesolimbic reward system and social behaviour network.* J Comp Neurol, 519, 3599-3639 (C4).

Ostojic, L et al (2013). *Evidence suggesting that desire-state attribution may govern food sharing in Eurasian jays.* PNAS, 110, 4123-4128 (C6).

Patel, AD et al (2009). *Experimental evidence for synchronization to a musical beat in a nonhuman animal.* Curr Biol, 19, 827-830 (C3).

Paz-y-Mino, GC et al (2004). *Pinyon jays use transitive inference to predict social dominance.* Nature, 430, 778-781 (C4).

Pepperberg, IM (2002). *Cognitive and communicative abilities of grey parrots.* Curr Dir Psychol Sci, 11, 83-87 (C3).

Petkov, CI & Jarvis, ED (2012). *Birds, primates, and spoken language origins.* Front Evol Neurosci, 4, 12 (C7).

Prather, JF et al (2008). *Precise auditory-vocal mirroring in neurons for learned vocal communication.* Nature, 451, 305-310 (C3).

Prior, H et al (2008). *Mirror-induced behaviour in the magpie.* PLoS Biol, 6, e202 (C6).

Raby, CR et al (2007). *Planning for the future by western scrub-jays.* Nature, 445, 919-921 (C6).

Rogers, LJ et al (2004). *Advantages of having a lateralized brain.* Proc Roy Soc B: Biol Lett, 271, S420-S422 (C1).

Rugani, R et al (2009). *Arithmetic in newborn chicks.* Proc Roy Soc B, 276, 2451-2460 (C7).

Scheiber, IBR et al (2005). *Active and passive social support in families of greylag geese.* Behaviour, 142, 1535-1557 (C4).

Seed, AM et al (2006). *Investigating physical cognition in rooks, Corvus frugilegus.* Curr Biol, 16, 697-701 (C5).

Seed, AM et al (2007). *Postconflict third-party affiliation in rooks.* Curr Biol, 17, 152-158 (C4).

Seed, AM et al (2008). *Cooperative problem solving in rooks.* Proc Roy Soc B, 275, 1421-1429 (C4).

Seed, A et al (2009). *Intelligence in corvids and apes.* Ethology, 115, 401-420 (C7).

Shimizu, T & Bowers, AN (1999). *Visual circuits of the avian telencephalon: evolutionary implications.* Behav Brain Res, 98, 183-191 (C1).

Smirnova, A et al (2015). *Crows spontaneously exhibit analogical reasoning.* Curr Biol, 25, 256-260 (C7).

Taylor, AH et al (2007). *Spontaneous metatool use by New Caledonian crows.* Curr Biol, 17, 1504-1507 (C5).

Taylor, AH et al (2009). *Do New Caledonian crows solve physical problems through causal reasoning?* Proc Roy Soc B, 276, 247-254 (C5).

Tebbich, S et al (2001). *Do woodpecker finches acquire tool-use by social learning?* Proc Roy Soc B, 268, 2189-2193 (C5).

Tebbich, S et al (2002). *The ecology of tool-use in the woodpecker finch.* Ecol Lett, 5, 656-664 (C5).

Tebbich, S et al (2007). *Non-tool-using rooks solve the trap-tube problem.* Anim Cogn, 10, 225-231 (C5).

Templeton, CN et al (2005). *Allometry of alarm calls: Black-capped chickadees encode information about predator size.* Science, 308, 1934-1937 (C3).

Teschke, I & Tebbich, S (2011). *Physical cognition and tool use: performance of Darwin's finches in the two-trap tube task.* Anim Cogn, 14, 555-563 (C5).

Vander Wall, SB (1982). *An experimental analysis of cache recovery in Clark's nutcracker.* Anim Behav, 30, 84-94 (C2).

von Bayern, AMP & Emery, NJ (2009a). *Jackdaws respond to human attentional states and communicative cues in different contexts.* Curr Biol, 19, 602-606 (C3).

von Bayern, AMP & Emery, NJ (2009b). *Bonding, mentalizing and rationality.* In: Watanabe, S (Ed.) *Irrational Humans, Rational Animals.* Keio University Press: Tokyo (C3).

Watanabe, S et al (1995). *Pigeon's discrimination of paintings by Monet and Picasso.* J Exp Analysis Behav, 63, 165-174 (C7).

Weir, AAS et al (2002). *Shaping of hooks in New Caledonian crows.* Science, 297, 981 (C5).

Wiltschko, W & Wiltschko, R (1972). *Magnetic compass of European robins.* Science, 176, 62-64 (C2).

Wimpenny, JH et al (2009). *Cognitive processes associated with sequential tool use in New Caledonian crows.* PLoS ONE, 4, e6471 (C5).

FURTHER READING

Birkhead, T. (2012). *Bird Sense: What it's like to be a bird.* Bloomsbury: London.

Boehner, B. (2004). *Parrot Culture: Our 2500 year-long fascination with the world's most talkative bird.* University of Pennsylvania Press: Philadelphia.

de Waal, F. B. M. (2016). *Are We Smart Enough to Know How Smart Animals Are?* W. W. Norton & Co., New York.

Emery, N. (2006). *Cognitive ornithology: the evolution of avian intelligence.* Philosophical Transactions of the Royal Society B, 361, 23-43.

Emery, N. and Clayton, N. (2004). *The mentality of crows: Convergent evolution of intelligence in corvids and apes.* Science, 306, 1903-1907.

Heinrich, B. (1999). *Mind of the Raven.* Harper Collins Publishers: New York.

Hansell, M. (2007). *Built By Animals: The natural history of animal architecture.* Oxford University Press: Oxford.

Marzluff, J. and Angell, T. (2012). *Gifts of the Crow: How perception, emotion, and thought allow smart birds to behave like humans.* Free Press: New York.

Morell, V. (2013). *Animal Wise: The thoughts and emotions of our fellow creatures.* Crown Publishers: New York.

Pepperberg, I. (1999). *The Alex Studies: Cognitive and communicative abilities of grey parrots.* Harvard University Press: Cambridge, MA.

Savage, C. (1997). *Bird Brains: Intelligence of crows, ravens, magpies and jays.* Greystone Books: Canada.

Tudge, C. (2008). *Consider the Birds: Who they are and what they do.* Allen Lane: London.

INDEX